Henry Eoghan O'Brien
D.S.O., B.Sc., M.A., M.I.C.E., M.I.E.E., M.I.Loco.E., M.I.C.E.I.

An Engineer of Nobility

by
Gerald M. Beesley
B.Sc., C.Eng., F.I.Mech.E., F.I.E.I., F.I.R.S.E., F.P.W.I., F.I.A.E.

Published 2018 by the Author

First Edition

Text Copyright © Gerald Beesley, 2018

Illustrations © Various, as acknowledged in captions

All rights reserved. No part of this publication may be reproduced, stored in a retrieval system or transmitted in any form or by any means, electronic, mechanical, photocopying, scanning, recording or otherwise, without the prior written permission of the copyright owners.

The author has asserted his right under the
Copyright, Design and Patents Act, 1988,
to be identified as the author of this work

Designed and printed by Creative Print & Design, Wexford

ISBN 978-1-78926-677-1   2018/500

Table of Contents

	Foreword	3
	Preamble	4
	Introduction and Acknowledgements	6
Chapter One	The O'Brien Clan	9
Chapter Two	Railway Relations	16
Chapter Three	Dromoland to Killiney	28
Chapter Four	The Formative Years	40
Chapter Five	Liverpool - Southport Electrification	62
Chapter Six	Life with an Electric Railway	80
Chapter Seven	Management of Workshops	96
Chapter Eight	With the Royal Engineers	119
Chapter Nine	National Projectile Factories	127
Chapter Ten	Light Railways Directorate	143
Chapter Eleven	Assigned to two Ministries	158
Chapter Twelve	Return to Horwich	172
Chapter Thirteen	Retirement at Killiney	187
Appendix One	L&YR Chief Mechanical Engineer's department officers	209
Appendix Two	L&YR Electric Stock (Liverpool and Manchester lines)	210
Appendix Three	Carriage orders completed at Newton Heath 1909-10	211
Appendix Four	Wagon types built at Newton Heath 1909-10	212
Appendix Five	Locomotive orders completed at Horwich 1910-21	212
Appendix Six	Locomotives built at Horwich 1910-21	213
Appendix Seven	Progress of National projectile Factories, 15th April 1916	214
Appendix Eight	Summary of Up Trains, Deir Sineid – Junction Station 21st November to 7th December 1917	215
Appendix Nine	Electric Traction – paper read by Col. H.E. O'Brien to the Irish Centre of the I.E.E., Dublin, 15th January 1931	216
Appendix Ten	An Alpine Adventure	224
Appendix Eleven	Fishing Recollections	225
	Bibliography	226
	Index	231

Official portrait of Colonel H.E. O'Brien taken on the occasion of his appointment in January 1922 as Electrical Engineer of the London & North Western Railway - [Photo ©The National Archives, Kew]

Foreword

I remember Eoghan O'Brien well, for he did not die until I was 28. He was my beloved grandfather, the more beloved because my father was killed before I was two, and so he took his place in many ways. For example, as a small child watching him shave, I too used to cover my face with shaving cream and scrape it off, with a bladeless razor. Granny, always known to Eoghan as Francesca, with the Italian pronunciation, was a less intense figure, though she introduced me to ponies and riding. Because I was my father's only child, and he theirs, I always spent most of the summer with them at *Mount Eagle*, Killiney, though for steadily shorter visits once I had gone to university.

But Grandpapa I could talk to; could discuss things with him. In the mornings, he lay in bed, listening to Swiss Radio, not Radio Éireann, not even the BBC, and he talked about the news. He talked to me too about the railway that ran below the foot of the garden at *Mount Eagle*, a railway over which we had a private bridge, which we used to go down to the strand at the White Rock, a splendid place to bathe. Looking along the coast towards Bray Head, when the sun shone just right, one could see through the railway tunnel that pierced the hill. I knew he was an engineer, as my father had been, and he taught me to be interested in crawling under the car and getting oily, seeing the hidden but essential parts.

From as early as I can remember he encouraged me to scramble up the boulders and mini-cliffs on Killiney Hill rather than walking round them; when I was a bit older he took me to Dalkey Quarry, on the other side of Telegraph Hill, where there were graded climbs, and I met people who knew him from the Irish Mountaineering Club. Twice he took me to the Alps, to Saas Fee, and to Arolla, where we walked on the lower slopes among the flowers, and took a guide and a porter – for he was old and I was young – to climb the summits.

He also introduced me to fishing, sea fishing for mackerel in Dalkey Sound, and then, as I got a little more adept at manoeuvring a rod, to casting a fly for trout, and in due course for salmon. Mostly we went to fish the Slaney at Huntington Castle, but one year he took me to the Dee.

Even to the end of his life he remained interested in things and people. By the end, and he died at 91, he would say 'Tired now' and just switch off his attention and fall asleep, but when he awoke he was alert and intelligent. My father's death was a deep sadness for him, and my grandmother's dementia another, but he felt he had had a good life, and was always pleased to hear about his great-grandchildren, and my academic life.

Olivia F. Robinson, FRSE, FRHistS, kMOestAk
Professor Emeritus, School of Law, University of Glasgow

Preamble

Research undertaken for a history of the Dublin & South Eastern Railway[1] (D&SER) unearthed a company file relating to various proposals for electrification of the line between Amiens Street (now Connolly) station, Dublin, and Bray. Amongst other papers, the file contains correspondence, dated June 1920, between George Wild, Locomotive Engineer of the D&SER, and E. O'Brien, Chief Mechanical Engineer at the Ministry of Transport in Dublin. At that time the wider historic relevance of these particular documents was not fully appreciated.

Subsequent research into traction experiments with the Drumm Battery, carried out on the Great Southern Railways (GSR) in Ireland during the 1930s, turned up files of correspondence between Col. Eoghan O'Brien of *Mount Eagle*, Killiney, Co. Dublin and W.H. Morton, Chief Mechanical Engineer and later General Manager of the GSR. On the basis of identical signatures, it was deduced that these and the earlier letters were from the same person. The question then arose; who was Col. Eoghan O'Brien?

The author was aware that between 1903 and 1922 one of the most important souls on the Lancashire & Yorkshire Railway (L&YR) was H.E. O'Brien; an Irishman from a family with means, about whom very little had been written. Along with a number of other distinguished railway engineers, including Sir Henry Fowler, Sir Nigel Gresley, and Richard Maunsell, he was ranked as one of the really brilliant pupils to emerge from Horwich under the training programme for engineers initiated by John Aspinall, Chief Mechanical Engineer of the L&YR. Was this H.E. O'Brien the same person as the correspondent who had been discovered in connection with Irish railway matters?

Relevant editions of *Burke's Peerage & Baronetage* and *Who's Who* provided the answer; readily confirming that, prior to his retirement from the London Midland & Scottish Railway (LMS) in 1925, Henry Eoghan O'Brien, of *Mount Eagle*, Killiney, had indeed been Assistant Chief Mechanical Engineer on the L&YR. He had subsequently been appointed Electrical Engineer on the London & North Western Railway following its merger with the L&YR in January 1922; a position he retained just one year later when the railway grouping of 1923 resulted in the formation of the LMS. Service during the First World War with the Royal Engineers and at the Ministry of Munitions saw him rise to the rank of Colonel. He was twice mentioned in dispatches and was awarded the DSO in the New Year's Honours of January 1918.

---

1 E. Shepherd & G. Beesley; *The Dublin & South Eastern Railway — an illustrated history*, Midland Publishing Limited, Leicester, 1998.

In this volume the author has endeavoured to present the biography of a railway engineer in whom Aspinall saw so bright a future that he promoted him at the young age of 26 to be Resident Engineer on the electrification project for the Liverpool, Crosby and Southport line — a major undertaking and one of the first two mainline electrification schemes to be brought into service on railways in the U.K[2]. As Assistant Chief Mechanical Engineer, and the leading electrical expert on the L&YR, H.E. O'Brien's career would appear to have had no limitations, but for the differences of opinion that arose on the LMS in regard to long-term motive power policy. This book is devoted as much to the contemporary railway scene with which H.E. O'Brien was involved as to the man himself, and it is hoped that this approach adequately demonstrates his all-round capability as one of the more important railway engineers.

H.E. O'Brien's first major achievement was as Resident Engineer for the electrification of the Liverpool, Southport and Crossens section of the Lancashire & Yorkshire Railway. This view shows one of the standard L&YR 4-car Liverpool–Southport sets in service - [HOR-F-225 © NRM, York]

---

2 Electric services commenced in 1904 on the L&YR (22nd March) and on the North Eastern Railway (29th March).

Introduction and Acknowledgements

The year 1919 saw Frederick Walter Beesley complete his schooling at the Edinburgh Academy and move to Glasgow where, on 6th April 1920 he entered the service of the Caledonian Railway (CR) at their St. Rollox Works as a premium apprentice of William Pickersgill, Chief Mechanical Engineer. Not long after he joined the CR, the Institution of Locomotive Engineers (I.Loco.E.) formed a Scottish Centre, holding its inaugural meeting in the Societies' Room of the Royal Technical College, Glasgow, on Friday 29th October 1920, with Pickersgill, President of the I.Loco.E., in the chair.

No doubt encouraged by Pickersgill, the young Fred Beesley enrolled in the I.Loco.E. as a student member, and it is quite possible that he was amongst those members who assembled at 7.30 p.m. on 26th November 1920 for the first ordinary meeting of the Scottish Centre; the Chairman of the Centre, Robert H. Whitelegg, Chief Mechanical Engineer of the Glasgow & South Western Railway, presiding. There, in the company of the Scottish locomotive engineers of the day, Fred Beesley would have listened to a paper on *The Management of a Locomotive Repair Shop* read by Col. H.E. O'Brien[3], Assistant Chief Mechanical Engineer of the Lancashire & Yorkshire Railway, and he would have heard contributions to the discussion on the paper from Walter Chalmers, Chief Mechanical Engineer of the North British Railway and Irvine Kempt, Assistant Locomotive Superintendent on his own railway, the Caledonian.

Fred Beesley moved to Dublin in April 1946 where he was to reside within walking distance of Sandymount station, which was on the former D&SER line to Bray; just 8¼ miles distant from Killiney where Col. O'Brien lived in retirement. Sandymount station, and the trains that operated on what had become the DSE section of Córas Iompair Éireann (CIÉ), were to provide the stimulus for a life-long interest in railways (as a career and as a railway engineering historian) for Fred Beesley's son; the author of this volume. It is unlikely that Fred Beesley met Col. O'Brien in Dublin, although, as Resident Engineer in Ireland for Babcock & Wilcox he would have come across many who shared the same professional background. However, it is interesting to consider that, 98 years prior to this biography being completed; Fred Beesley may well have listened to Col. O'Brien as he read his paper to the I.Loco.E. in Glasgow.[4]

The compilation of this biography would not have been possible without the generous help of the many people who made information available. First of all,

---

3 Col. H.E. O'Brien, *The Management of a Locomotive Repair Shop*; I.Loco.E. Journal, Vol.45, 1920.

4 As a Graduate member of the I.Loco.E. Fred Beesley also read a paper before the Scottish Centre, at Glasgow on 19th January 1922, for which R.H. Whitelegg presented him with a Silver Medal given by David Smith, Chief Draughtsman of the Glasgow & South Western Railway. (F.W. Beesley, *The Modern British Express Passenger Locomotive*; I.Loco.E. Journal, Vol.53, 1922.)

my sincere thanks must go to Prof. Olivia F. Robinson, granddaughter of Col. H.E. O'Brien, who not only encouraged my efforts but also shared her own recollections and provided material from family archives. Regarding the wider family circle, Raymond Refaussé and the staff at the Church of Ireland Representative Body Library, Rathgar, kindly assisted with biographical details about Revd Henry O'Brien, and the Waller, Lloyd and Maunsell clergy; while Dick Bullock of Melbourne provided much guidance in connection with the railway careers of Donatus O'Brien and Capt. William O'Brien.

For details of H.E. O'Brien's schooldays I am indebted to Jackie Tarrant-Barton, of the Old Etonians Association. The Secretary's department of CIÉ kindly allowed me access to the Board Minute books, Chief Engineers' reports and Locomotive Superintendents' letter books of the Dublin Wicklow & Wexford Railway (DW&WR) in order that I might obtain information of events that occurred on that railway during H.E. O'Brien's time with the company.

As always, the Irish Railway Record Society (IRRS) proved invaluable when it came to Irish railway matters, unearthing files containing letters from H.E. O'Brien when he was at the Irish Branch of the Ministry of Transport, Dublin, in 1920, and also (subsequent to his retirement) correspondence with W.H. Morton of the GSR. The Clements collection, held by the IRRS, also includes some papers bequeathed to the society by H.E. O'Brien. For access to all of this material I am particularly grateful to the Society's Hon. Librarian, Tim Moriarty, and late Archivist, Brendan Pender. Additional information on correspondence from H.E. O'Brien whilst at the Ministry of Transport in Dublin, in regard to matters on the Belfast & County Down Railway and Great Northern Railway (Ireland), was kindly provided by the late Des Coakham.

The National Archives (formerly PRO), Kew, was the source of much material on H.E. O'Brien's days on the L&YR and the LMS. The staff at Kew also assisted in regard to his military service and, for additional information covering that period, I must thank Craig Bowen, Registrar and Assistant Curator, Royal Engineers' Museum, Chatham, and Colin Harris, Modern Papers section, Bodleian Library, Oxford. Brian Donnelly, Irish National Archives, Dublin, assisted with material on the provincial office of the Ministry of Transport in Ireland and files in connection with the Drumm Battery experiments. The National Railway Museum, York, was a source of much useful data and photographs, and the assistance of Philip Atkins is gratefully acknowledged. Dudley Fowkes, Midland Railway Trust, Ripley, kindly provided access to the letter books of George Hughes, Chief Mechanical Engineer of the L&YR, which threw light on many matters with which H.E. O'Brien was concerned.

I must also acknowledge the assistance of the respective librarians and their staffs at those three great bodies of professional engineering: the Institution of Civil Engineers, the Institution of Mechanical Engineers, and the Institution of Engineering and Technology (I.E.T) formerly the Institution of Electrical Engineers (I.E.E.). I am especially grateful to Anne Locker, Assistant Archivist at the I.E.T., for her unstinting efforts in responding to my persistent inquiries. The librarian of the Institution of Engineers of Ireland, John Callanan, also deserves a word of thanks for allowing me access to the records of the Irish Centre of the I.E.E. The staff of the Berkeley Library at Trinity College Dublin assisted with information in relation to H.E. O'Brien's time in the Engineering School as Lecturer in Railway Traction.

Thanks are also due to all those who have supplied the photographs that grace the pages of this work, especially to Ciarán Cooney, Hon. Photographic Archivist, IRRS, who not only provided images from the Society's collection but also enhanced the quality of a number of the other images. A very special word of thanks is also due to Robin Barnes who has graciously allowed me to reproduce his painting of H.E. O'Brien's proposed 2-Do-2 electric locomotive. Wherever possible, original photographers or copyright holders have been individually acknowledged. Finally, thanks must go to my good friend Ernie Shepherd, who read the manuscript and provided valuable critical advice, and to Noel Moore of Creative Print & Design, Wexford, for his work on the layout design and printing.

A work of this kind is only as good as the information that has been discovered at the time of writing. If, through oversight, I have failed to appropriately acknowledge any name or source I can only apologise for any such unintentional omission. Any errors of fact are, of course, my sole responsibility.

Gerald Beesley
Saltmills
New Ross
Co. Wexford
Y34 HP28

November, 2018

The O'Brien Clan

This is the land of Swift and Wilde,
of Yeats and Beckett,
of O'Casey and O'Connor,
and more than one O'Brien.

In the first few centuries of the Christian era Britain was invaded, first by the Romans and later by the Anglo-Saxons, but Ireland remained largely unaffected until near the end of the 8th century when the Vikings invaded it. At that time Ireland was divided into no less than 150 small 'kingdoms' called 'Tuatha'. Over the years these became grouped together into larger provincial kingdoms. The rulers of the five 'Fifths' — the provinces of Ulster, Munster, Meath, Leinster, and Connaught — were looked on as the principal kings in Ireland. Later, with consolidation of O'Neill power as high kings of Ireland, the smaller kingdom of Meath was assumed into Leinster, and Munster, the largest of the provinces, was divided into two kingdoms. By the twelfth century the division of Munster was formalised in the Desmond kingdom of the McCarthys in the south and the Thomond kingdom of the O'Briens in the north. The kingdom of Thomond covered an area approximating to the present counties of Clare and Limerick, and part of Tipperary.

Brian Boru, High King (Ard Rí) of all Ireland, 1002-14 - [Author's collection]

The royal line of Thomond descended directly from Brian Boru (or Boroimhe), who was born about 941 and reigned from the royal seat of his ancestors at *Kincora*, where the town of Killaloe, Co. Clare, now stands. In 960 Brian's brother, Mahoun, became King of North Munster. At that time Munster was being plagued by Viking raids and Brian managed to unite the province's rival royal clans, the northern Delcassians (to which he and Mahoun belonged) and the southern Eoghanists (descendants of Eoghan Mór), and led them to victory against Viking raiders from Limerick. The Eoghanists grew jealous of Delcassian power and formed an alliance with the Vikings. King Mahoun was treacherously murdered and Brian declared war on the Eoghanists and their Viking allies, defeating them and by 978 had extended his authority to the whole of Munster. Six years later, Brian was in control of the southern half of Ireland and by marriage into the Viking royal family in Dublin his power grew. He established himself as High King (Ard Rí) of all Ireland in 1002, thus brining to an end 600 years of dominance by the northern O'Neill clan. Family feuds led to the King of Leinster convincing the Vikings to join him in declaring war on Brian. The two sides met at the battle of Clontarf on

Good Friday 23rd April 1014 and, although his army was victorious, Brian fell by the sword of Brodar, the Danish admiral. His remains were escorted to Armagh Cathedral for burial, where they rest to this day.

From Brian there stemmed a sixteen-generation line of kings of North Munster and Thomond, spanning a period of 529 years after his death. The O'Brien clan remained major contestants for the high kingship until 1103 and, except for two short breaks, they also ruled over all of Munster until 1194, Donal Mór being the last of the clan to hold that title. On his accession in 1168 Donal Mór gifted to the Church a palace in Limerick that had been the seat of government in the early medieval city. St. Mary's cathedral was founded on the site and is recognised as one of the oldest buildings in the city. Parts of the palace, including the West Door, which tradition has it was the main entrance to the palace, are incorporated in the present structure of the Cathedral.

Cairbreach O'Brien, eldest son of Donal Mór, was the first of the family to formally assume the surname O'Brien; indicating his lineage was that 'of Brian'. The 57th and last King of Thomond was Murrough O'Brien who acceded in 1540. When Henry VIII of England decided to proclaim himself King of Ireland and trade English titles for traditional Gaelic authority, Murrough submitted his authority to the King. In return Murrough O'Brien was created Earl of Thomond, for life, and Baron Inchiquin not only for himself but also for his descendants. The formal proceedings took place on 1st July 1543 at Greenwich by the Thames, with Murrough's nephew Donough Ramhar O'Brien, who was a minor, also in attendance.

In 1551, on the death of Murrough O'Brien, the 1st Earl of Thomond, the Earldom passed to his nephew, Donough Ramhar O'Brien and thence to Donough Ramhar's descendants; the Barony passing at the same time to Murrough's eldest son, Dermod O'Brien. Murrough's will contains the first recorded reference to *Lemeneagh* ('leap of the deer')[5] and *Dromoland* (variously translated as 'hill of litigation' or 'hill of the lake'). These estates were assigned to Murrough's youngest son, Donough, and thereafter this branch of the family became known as the Lemeneagh or Dromoland O'Briens. *Lemeneagh* was a fortified manor house consisting of two combined structures. The major part, the manor house on the eastern end, was built onto the original five-storey tower house dating from 1480. It is thought to have been built by Donough's great grandson, Conor O'Brien, on the foundations of an earlier hall, after he inherited the estate in 1637. In addition to extending the residence he surrounded the estate with a fortified wall and an impressive front entrance gate, which displayed the inscription '**Built in the year 1643 by Conor O'Brien and by Mary ní Mahon, wife of the said Conor**'. Conor's son, Donat, improved the property and, in addition to work on the gardens, a canal was added and the carriage drive was lined with trees, but in 1705 Donat left Lemeneagh for

---
5 Lemeneagh is on the road from Corofin to Kilfenora in County Clare.

CHAPTER ONE **The O'Brien Clan**

Royal lineage of Thomond

Brian Boru
King of Munster 978 – 1014
High King of Ireland 1002 - 14

Teige
King of North Munster 1014 -23

Donough
King of Munster 1023 – 64

Turlogh Mór
King of Munster 1064 - 86
High King of Ireland 1072 - 86

Murtogh
King of Munster 1086 - 1116
High King of Ireland 1086 - 1116

Dermod
King of Munster 1116 - 18

Conor
King of Thomond 1118 - 42

Turlogh
King of Thomond & Munster 1142 - 67

Murtogh
King of Thomond 1167 - 68

Donal Mór
King of Thomond & Munster 1168 - 94

Murtogh Fionn
King of Thomond 1194 - 98

Conor Rua
King of Thomond 1198 - 1210

Donough Cairbreach
King of Thomond 1210 - 42

Conor na Suidane
King of Thomond 1242 - 68

Teige Caol Uisce
d. 1259

Brian Rua
King of Thomond 1268 - 77

Turlogh Mór
King of Thomond 1277 - 1306

Donal
d. 1280

Donough
King of Thomond 1306 - 11

Murtogh
King of Thomond 1313 - 43

Dermod
King of Thomond 1311 - 13

Brian Bán
King of Thomond 1343 - 50

Dermod
King of Thomond 1350 - 60

Mahon Moinmoy
King of Thomond 1360 - 69

Turlogh Maol
King of Thomond 1375 - 1398

Brian Catha an Aonaigh
King of Thomond 1369 - 75

Conor
King of Thomond 1398 - 1426

Teige na Gaoidh Mór
King of Thomond 1426 - 38

Mahon Doll
King of Thomond 1438 - 44

Turlogh Beg
King of Thomond 1444 - 59

Teige an Chomhard
King of Thomond 1459 - 66

Conor Mór na Shrona
King of Thomond 1466 - 96

Turlogh Óg
King of Thomond 1496 - 99

Turlogh Donn
King of Thomond 1499 – 1528

Dromoland and thereafter the castle fell into decay.

The present-day *Dromoland Castle*, near Newmarket-on-Fergus, County Clare, is the third structure to stand on the site. The original castle was a 15th century tower house, and passed through various hands between 1551 and 1684. The aforementioned Donough O'Brien was in residence for over 30 years but, after he was hanged in Limerick in 1582 on charges of rebellion, all of his property was forfeited to the crown. The High Sheriff for the county, Sir George Cusack, took possession, but when he was killed some years later various O'Briens then attempted to re-possess *Dromoland*. The 4th Earl of Thomond claimed sole ownership and endeavoured to exclude Donough's son, Conor O'Brien, but this dispute does not appear to have been resolved, for when Conor died in 1603 he left *Dromoland* to his son, another Donough, who was only about eight years old at that time. A legal battle ensued between the 4th Earl and Conor's widow, which was settled by arbitration in 1613, the 4th Earl becoming owner of *Dromoland* on payment of compensation to Conor's widow. When Donough was older he refused to accept this agreement and, after the 4th Earl's death in 1624, he continued to pursue his claim through the Court of Wards and Liveries in Dublin. Donough was granted entry "on all the manors, lands and tenements of his late father" on payment of a fine in 1629. *Dromoland* was not listed among the many properties named in this settlement and so its ownership remained with the Earls of Thomond for another 55 years, although the 5th Earl did transfer two other properties to Donough as compensation.

The 4th Earl leased *Dromoland* to William Starkey, whose son Robert was there when the rebellion of 1641 began. It seems that Robert Starkey either fled the area or sublet the estate for; in 1642, the aforementioned Conor O'Brien of *Lemeneagh* seized the castle, thereby continuing his father's claim to *Dromoland*. Conor, who was married to Máire Rua[6], was killed at the Inchicronan ambush in 1651, but when Cromwellian troops brought his body to *Lemeneagh* his wife would not allow it to be brought into the castle and denied any knowledge of him. "We want no dead men around here", she proclaimed, for Máire Rua was well aware that forfeiture of the family estates would be the penalty for Connor's rebellion. She rode into Limerick, demanded an audience with General Ireton, and offered to marry any officer of his choosing.[7] John Cooper, a fifth commissioned officer in the cavalry, obliged, and was accepted. Within a few days of Conor's death Máire Rua had married the Cromwellian soldier, thus ensuring that the O'Brien estates were preserved at the general confiscation.

Although Máire Rua had secured the *Lemeneagh* and *Cratloe* estates, the problem of *Dromoland* had still not been resolved, and Conor and Máire Rua's eldest son, the

---
6 Mary ní (daughter of) Mahon, or Red Mary as she was known in recognition of her flaming red hair.
7 General Henry Ireton (1611–51) was the son-in-law of Oliver Cromwell.

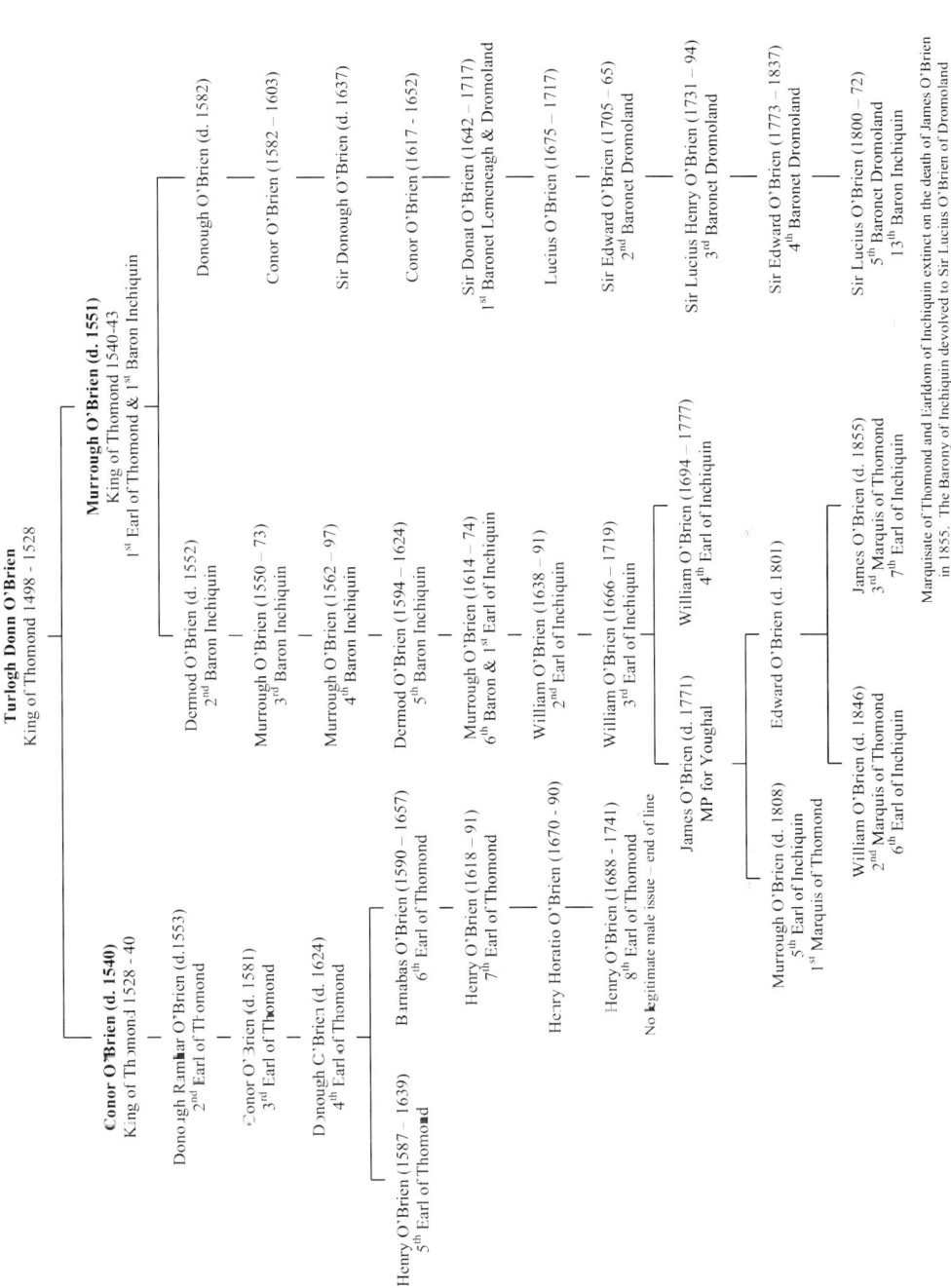

Barony of Inchiquin

Turlogh Donn O'Brien
King of Thomond 1498 - 1528

- **Conor O'Brien (d. 1540)**
  King of Thomond 1528 - 40
  - Donough Ramhar O'Brien (d.1553)
    2nd Earl of Thomond
  - Conor O'Brien (d. 1581)
    3rd Earl of Thomond
  - Donough C'Brien (d. 1624)
    4th Earl of Thomond
    - Henry O'Brien (1587 – 1639)
      5th Earl of Thomond
      - Barnabas O'Brien (1590 – 1657)
        6th Earl of Thomond
        - Henry O'Brien (1618 – 91)
          7th Earl of Thomond
          - Henry Horatio O'Brien (1670 – 90)
          - Henry O'Brien (1688 - 1741)
            8th Earl of Thomond
            No legitimate male issue – end of line

- **Murrough O'Brien (d. 1551)**
  King of Thomond 1540-43
  1st Earl of Thomond & 1st Baron Inchiquin
  - Dermod O'Brien (d. 1552)
    2nd Baron Inchiquin
    - Murrough O'Brien (1550 – 73)
      3rd Baron Inchiquin
      - Murrough O'Brien (1562 – 97)
        4th Baron Inchiquin
        - Dermod O'Brien (1594 – 1624)
          5th Baron Inchiquin
          - Murrough O'Brien (1614 – 74)
            6th Baron & 1st Earl of Inchiquin
            - William O'Brien (1638 – 91)
              2nd Earl of Inchiquin
              - William O'Brien (1666 – 1719)
                3rd Earl of Inchiquin
                - William O'Brien (1694 – 1777)
                  4th Earl of Inchiquin
                  - James O'Brien (d. 1771)
                    MP for Youghal
                    - Edward O'Brien (d. 1801)
                      - Murrough O'Brien (d. 1808)
                        5th Earl of Inchiquin
                        1st Marquis of Thomond
                        - William O'Brien (d. 1846)
                          2nd Marquis of Thomond
                          6th Earl of Inchiquin
                          - James O'Brien (d. 1855)
                            3rd Marquis of Thomond
                            7th Earl of Inchiquin
  - Donough O'Brien (d. 1582)
    - Conor O'Brien (1582 – 1603)
      - Sir Donough O'Brien (d. 1637)
        - Conor O'Brien (1617 - 1652)
          - Sir Donat O'Brien (1642 – 1717)
            1st Baronet Lemeneagh & Dromoland
            - Lucius O'Brien (1675 – 1717)
              - Sir Edward O'Brien (1705 – 65)
                2nd Baronet Dromoland
                - Sir Lucius Henry O'Brien (1731 – 94)
                  3rd Baronet Dromoland
                  - Sir Edward O'Brien (1773 – 1837)
                    4th Baronet Dromoland
                    - Sir Lucius O'Brien (1800 – 72)
                      5th Baronet Dromoland
                      13th Baron Inchiquin

Marquisate of Thomond and Earldom of Inchiquin extinct on the death of James O'Brien in 1855. The Barony of Inchiquin devolved to Sir Lucius O'Brien of Dromoland young Donat O'Brien, became heir to the family claim. However, Robert Starkey resumed the lease, and in 1666 *Dromoland* was sub-let to Col. Daniel O'Brien from Carrigaholt, 3rd Viscount Clare; a grandson of the 4th Earl of Thomond's brother. Finally, in 1684, the 7th Earl of Thomond assigned the freehold of the castle to Donat O'Brien; the lands of *Dromoland* being leased in perpetuity for a yearly rent of £117. About 1713-14 a new wing was added onto the old tower house; later in the 18th century a new house was built, based on a design from the Queen Anne period, and the original *Dromoland Castle* was pulled down. Sir Edward O'Brien, 4th Baronet Dromoland, decided upon the construction of a new castle in 1820. It was built alongside *Dromoland House* to a design in the Gothic-style by the brothers James and George Richard Pain and was completed in 1835. The two buildings stood together for a short while until the 18th century house was demolished.

Dromoland Castle, Co. Clare, seat of the Barons of Inchiquin -
[Lafayette]

The O'Brien family is one of the few native Irish houses to be found in the peerage of Ireland and, as we have seen, deduces its descent from the royal line of Thomond. The senior branches of the family were the Earls of Thomond and Earls of Inchiquin. Murrough O'Brien, 6th Baron Inchiquin, was made Earl of Inchiquin by Charles II in 1654, and in 1800 his descendant, another Murrough, the 5th Earl of Inchiquin, was created Marquis of Thomond. The Thomond earldom had previously become extinct in 1741 upon the death of Henry O'Brien, the 8th Earl. In July 1855, on the death of James, the 3rd Marquis of Thomond, both the marquisate of Thomond and the earldom of Inchiquin became extinct. The barony of Inchiquin, however, passed to Sir Lucius O'Brien, 5th Baronet Dromoland, who then became 13th Baron Inchiquin. *Dromoland Castle* continued to be the seat of the Inchiquin barony until 1962 when the castle and part of the estate were sold, the castle thereafter being transformed into one of Ireland's finest hotels.

To continue our story, we must return to the 17th century O'Briens of *Dromoland*.

Donat was twice married, firstly to Lucia Hamilton, by whom he had a son and heir, Lucius; born in 1675. Lucia died in the following year, probably in childbirth, and on 23rd July 1677 Donat married Elizabeth Gray, daughter of Major Joseph Deane and widow of the late Dr Henry Gray, by whom he had three children; a son, Henry, and two daughters. Sir Donat, who was created Baronet Lemeneagh and Dromoland on 9th November 1686, acquired *Stonehall*, in the parish of Kilcronry, Co. Clare, for his son Henry from Sir Henry Ingoldsby. Henry married Susannah Stafford, daughter and co-heiress of William Stafford of Blathewycke Park, Northamptonshire and they lived at *Stonehall* until the death of Susannah's father. The descendants of Henry of Stonehall included Stafford O'Brien (1783–1864), High Sheriff of Rutland, and Stafford O'Brien's son, Augustus Stafford O'Brien (1811-57), MP for Northamptonshire (1841-57) and First Secretary at the Admiralty (1852-53).

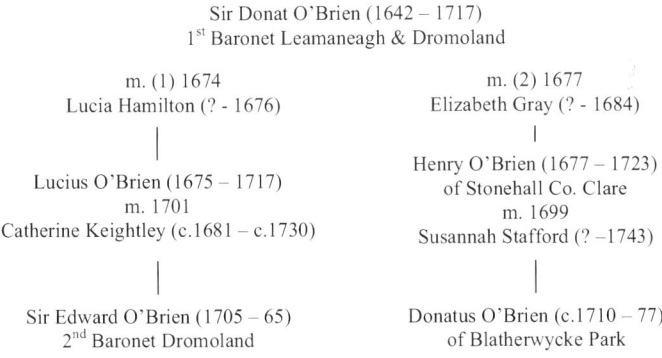

In the spring of 1701, Lucius O'Brien, Sir Donat's heir, married Catherine Keightley, a cousin of Queen Mary and Queen Anne[8], and moved into *Cratloe House*. They had two sons and two daughters, the eldest son Edward being born in 1705. Lucius O'Brien was in France when he fell ill and he died in Paris on 6th January 1717, and when Sir Donat died on 18th November in the same year, the title and *Dromoland* estate passed to the twelve year old Edward. Due to Sir Edward's young age, Catherine, with the help of several advisors, administered the estate. In 1726 Sir Edward married Mary Hickman and, with his coming of age, his mother handed over the running of *Dromoland* to him. Sir Edward's eldest son and heir, Lucius Henry, was born in London in 1731 and a year later a second son, Donough, arrived. However, before proceeding further with the Dromoland-lineage we must stay with the Cratloe O'Brien's for a moment, for it is in that branch of the family that we find the first O'Brien connections with railways.

---

8 Thomas Keightley, Catherine's father, was married to Lady Frances Hyde, sister of Lady Anne Hyde who was the wife of King James II and mother of Queen Mary II and Queen Anne. Lady Frances and Lady Anne were the daughters of Lord Clarendon.

Railway Relations

Cratloe House, originally called *Cratloe Hall*, was reputedly started in 1730. It is an impressive example of the Irish long house, a style that existed since medieval times. Set in its own grounds, it had been considerably extended by the middle of the nineteenth century and today enjoys the distinction of being the only surviving Irish long house that is still lived in by a family; the direct descendants of Lucius O'Brien. It is not known exactly when Sir Edward's son, Donough O'Brien, first resided at *Cratloe*, but in 1761 he married Mary Henn, a sister of William Henn of Paradise, Kildysart, Co. Clare, and was certainly there by 1765. Thirty years later Mary's niece, Elizabeth Henn, was to marry Bolton Waller of *Castletown* and *Shannon Grove*, Pallaskenry, Co. Limerick, eldest son of John Thomas Waller, High Sheriff of Limerick and his wife Elizabeth, daughter of Revd Richard Maunsell, Rector of Rathkeale, Co. Limerick. It was through this particular marriage that the first distant connection was to be established with the family of another famous railway engineer, Richard Edward Lloyd Maunsell, but we shall come to details of his other connections with the O'Briens a little later. Donough O'Brien's eldest son, Lucius, married Mary Callander and in 1797, on the death of his father, he inherited *Cratloe House* and estate. Our interest centres on two of Lucius' sons, William and Donatus.

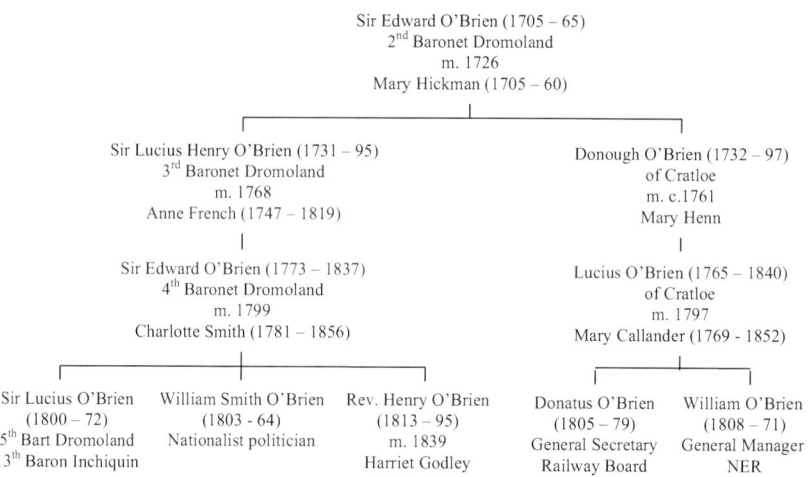

Donatus O'Brien, third son of Lucius O'Brien of *Cratloe House*, was born in 1805. He was a 4th cousin of the aforementioned Augustus Stafford O'Brien MP and a nephew of Sir James Robert George Graham, MP — First Lord of the Admiralty (1830-34 and 1853-55) and Home Secretary (1841-45).[9] His career commenced as a gentleman cadet at the Royal Military College, Sandhurst. After training he

---

9 Mary, née Callander, Donatus and William's mother, and Sir James' wife, Fanny, were half-sisters; daughters of Col. James Callander of Craigforth and Ardkinglass in Stirlingshire

served with the 20th and 21st Regiments of Foot and was promoted to Lieutenant with the 62nd Regiment of Foot from 5th February 1829. He was transferred to the Royal Staff Corps in May of the same year and promoted to the rank of Captain on 20th July 1838. He was married at St Mary Magdalene church, Woolwich, on 12th November 1835 to Elizabeth McCleverty, the only daughter of Colonel McCleverty, Commandant of the Woolwich Division of Royal Marines. Following the disbandment of the Royal Staff Corps he served with the Royal Engineers until retirement from the military in 1839.

Capt. Donatus O'Brien, RE, was appointed Secretary of the Great North of England Railway in July 1839, a post that he held until his appointment as private secretary to his uncle, Sir James Graham, MP in 1841[10]. He was elected a director of the Newcastle & Darlington Junction Railway Board at its first meeting on 7th July 1842, a post that he relinquished two years later on joining the Railway Board. His appointment as a member of the Railway Board came about in the following manner. On 19th July 1844, William Gladstone, the President of the Board of Trade (BoT), obtained a resolution in the House of Commons that gave him the necessary power to enlarge the Railway Department and place its functions under a distinct Railway Board. He advised Sir Robert Peel that the enlarged department should be staffed with persons who had practical experience in railway matters. The 4th Earl of Bessborough, deputy chairman of the Great Western Railway (GWR), was asked for an opinion, and one of the persons that he recommended was Donatus O'Brien.[11]

The members of the Railway Board, who were appointed by minute of 6th August 1844, were the Earl of Dalhousie (Chairman), Gen. Charles William Pasley (Inspector General of Railways), George Richardson Porter (Superintendent of the Railway Department), and Donatus O'Brien and Samuel Laing (Joint Secretaries[12]). In railway circles these gentlemen became known as the 'Five Kings'. An Assistant Inspector General of Railways, Capt. Joshua Coddington, was subsequently appointed. Office space was limited and Donatus had to share a room with Dalhousie's private secretary in the building on the corner of Whitehall and Downing Street that the BoT and Privy Council occupied.

The first meeting of the Railway Board was held at Council Chambers, Whitehall,

---

10 Other sources infer that Donatus was private secretary to Sir James Graham when he was First Lord of the Admiralty. This was not the case during 1830-34, when Sir James' private secretary was firstly Edward Stewart and secondly Major George Graham. Nor was it so for his second term as First Lord (1852-55) when his private secretary was Capt. Henry Higgins Donatus O'Brien, an elder brother of Donatus. In fact, it was while Sir James Graham was Home Secretary (1841-46) that Donatus served as his private secretary until appointed General Secretary to the Railway Board.

11 John William Ponsonby (1781-1847), 4th Earl of Bessborough, joined the GWR Board in 1839 and became deputy chairman in 1843. His sister was Lady Caroline Lamb.

12 Although appointed as Joint Secretaries, Samuel Laing was referred to as the Law Secretary and Donatus O'Brien as the General Secretary.

on 14th August 1844. Although the main purpose of the Board was to produce reports on the various railway Bills that were being proposed, they did attend to other matters of detail. One item that they became concerned about was the ever-increasing length of trains for pleasure excursions that were being adopted on many railways at that time. Donatus O'Brien was asked to investigate and his report on the safety of 'Monster Trains' was read at the Board meeting of 18th September 1844. At the same meeting consideration was resumed on Gen. Pasley's memorandum regarding new covered third class carriages that the GWR proposed to use, and it was directed that "the Inspector General and Mr O'Brien confer immediately with the directors of the GWR on the subject".[13] The same course was to be pursued with the Committee of the London & Brighton, London & Croydon and South Eastern railway companies and the results of the meetings were to be reported.

As the reports that were produced by the Railway Board normally recommend certain railway schemes at the expense of others, it was to be expected that the disappointed parties would object, particularly in view of the high value placed on the outcomes by the sponsors of the schemes. Speculative interests became inflamed and it was said that, in certain cases, the Board's decisions were known by some parties before they had been published in the *London Gazette*. In February 1845, it was suggested by a writer in the *Economist* that Donatus' brother, William, being at the time Secretary of the South Eastern Railway, had influenced the Board's view on schemes in which that company was interested, and that those 'in the know' had used such knowledge to their own advantage.

Sir James Graham, speaking in a House of Commons debate on 11th February, stated that it had been a condition of Donatus' appointment to the Board that any connection which he might then have had with any railway speculation should immediately cease, and that this requirement had been complied with. Sir James also suggested that the writer in the *Economist* may have come to some wrong conclusion on account of the identical surnames, and he assured the House that neither directly nor indirectly had there been any communication between the two brothers as to speculating in shares.[14] The Earl of Dalhousie, in laying the Railway Board's first report before the House of Lords on 13th February, indicated that Donatus, having foreseen the possibility of such insinuations, had formally abstained from taking any part in the consideration of the merits of the South Eastern Railway scheme. He also described the precautions which had been taken by the Board to prevent information of their decisions leaking out before they were announced.

Eventually, Dalhousie suggested that the Board should give up reporting on railway

---

13 National Archives (PRO) MT 13/1
14 Hansard's Parliamentary Debates, Vol. LXXVII, February 1845

Bills, unless asked to do so, and Peel agreed. The final meeting of the Railway Board took place on 23rd July 1845 at which "a copy of the Minute of Lords of the Committee of Privy Council for Trade dated 10th July 1845" was laid upon the table. [15] Dalhousie stated that before the proceedings of the Board were finally closed in pursuance of that Minute he was "desirous of expressing to the different members of the Board the sense entertained by the Lords of the Committee of the integrity, the diligence and ability with which they have discharged the onerous duties that have been imposed on them since the formation of the Railway Board in August 1844".[16] To this he added his own personal expression of thanks to the Board collectively and to each member of it individually for the cordial manner in which they had co-operated with him and for the ready and efficient aid which he had at all times received from them in conducting the business of the Department

The Lords of the Committee of Privy Council for Trade thereafter transacted railway business in the same manner as other ordinary business of their Committee, with the Earl of Dalhousie reporting to them on railway matters. Of the former members of the Railway Board, G.R. Porter and Gen. Pasley reverted to their previous positions in the Railway Department at the BoT; Samuel Laing, the Law Secretary, resigned and for the next year Donatus O'Brien alone dealt with all secretarial matters. As the chief non-technical officer and secretary to the Railway Department he remained in office until 12th November 1846, after which his career took him away from railways. In 1848, Donatus was appointed to the position of 2nd Inspector of Prisons, Midland & Eastern Districts, under Lieut-Col. Sir Joshua Jebb, Knight Surveyor General of Convict Prisons, and in the following year he was promoted to Director of Convict Prisons. He retired from the prison service in 1865 and was widowed in late 1870 when his wife, Elizabeth, died. He continued to live at 16 Gloucester Place, Hyde Park, until his death on 19th May 1879, not quite three years after the birth of H.E. O'Brien — his second cousin twice removed.

William O'Brien, fifth son of Lucius O'Brien was born at *Cratloe House* in 1808. His career also commenced at the Royal Military College, Sandhurst, as a gentleman cadet and, following training, he joined the Royal Staff Corps in 1826. He transferred to the 83rd Regiment in 1829 and was promoted to Lieutenant with the 53rd Regiment of Foot from 8th June 1830 and to Captain from 13th May 1836, serving with that Regiment until his retirement on 27th March 1840.

In 1841 William was appointed Secretary of the Great North of England Railway in succession to his brother Donatus, and in the same year he married a distant cousin, Katherine Lucy O'Brien, the granddaughter of James O'Brien, MP for Youghal. Whilst residing at Darlington their first child, Donatus, was born in June 1844. In November 1844 he moved south and briefly held the same position with the South

---

15 National Archives (PRO) MT 13/3
16 ibid

Eastern Railway. The charge of partiality laid at the Railway Board in February 1845 regarding decisions on schemes in which the South Eastern Railway was interested appears to have led to William's early departure from that company. However, in mid-1845 he joined the Wilts, Somerset & Weymouth Railway where he gained further experience under Charles Saunders, Secretary of the Great Western Railway and during that period his second child, a daughter, Kate, was born at Bradford-on-Avon in 1848.

Capt. William O'Brien, General Manager of the North Eastern Railway, 1854-71 - [G.B. Black]

Following the purchase of the Wilts, Somerset & Weymouth by the Great Western in March 1850, William returned to the north-east of England on his appointment as the Secretary of the York, Newcastle & Berwick Railway, which was the successor in title to the Great North of England Railway.[17] This move to Newcastle-upon-Tyne brought with it a new home at 19 Rye Hill, in the Elswick district, where two more daughters, Constance and Frances Elizabeth, were born in 1850 and 1853. When the York Newcastle & Berwick Railway amalgamated with the York & North Midland Railway and the Leeds Northern Railway on 31st July 1854 to form the North Eastern Railway (NER), William was confirmed as Secretary of the enlarged company. Very shortly thereafter, following the Bramhope Tunnel accident on 19th September 1854, Thomas Elliot Harrison (who up to that time had undertaken the dual role of General Manager and Chief Engineer) relinquished the former post. William O'Brien was appointed General Manager in his stead and also continued to act as Secretary until those two offices were separated in 1856. With the new role on the NER he moved to a residence at Blossom Hill Crescent, near The Mount in York.

William O'Brien soon settled down to tackle a task much greater than any other he had previously undertaken; the position he held requiring his involvement in many difficult matters. In 1855 he played a key role in drawing up the second Octuple Agreement, which apportioned the receipts arising from railway traffic between England and Scotland among the companies that were handling it. Much of the credit for the NER's prosperity throughout the 1860s must certainly have belonged to him for the successful manner in which he concluded that Agreement. Mid-way through 1857 tragedy struck the family when Katherine died, but William was to marry again in 1858. His second wife, Margaretta Burgoyne, was born in Belfast

---

17 The Newcastle & Darlington Junction Railway purchased the Great North of England Railway in 1846 and was re-titled the York & Newcastle Railway. In 1847 the York & Newcastle Railway merged with the Newcastle & Berwick Railway to form the York, Newcastle & Berwick Railway.

in 1817 and the union produced one child, Florence, who was born at Paddington, London, on 24th February 1860.

Another problem that William O'Brien dealt with was probably even more difficult, namely the amalgamation of the Newcastle & Carlisle Railway and the Stockton & Darlington Railway with the NER in 1862-3. Both companies were strongly opposed to the arrangement; nevertheless the amalgamation was sanctioned. His diplomacy must have been instrumental in securing this achievement, which was of great value in extending the NER's territorial monopoly. It was, however, with day-to-day railway operations that William O'Brien came to confront problems that were to lead to his eventual downfall. The approach that the NER adopted in addressing several key matters, whilst not in isolation so serious, was, as we shall see, to have disastrous consequences in combination.

Firstly, no serious or sustained effort appears to have been made to rationalise the diverse rules and regulations of the constituent companies of the NER. This was to prove fatal when, as a consequence of the NER enginemen's strike of 1867, the company ruthlessly dismissed 900 of its servants, resulting in a major loss of knowledge and discipline in train working skills that had enabled the NER to cope with the varied systems in operation. To these was added a reluctance to introduce block working and the interlocking of points and signals[18] (safety systems that nearly all the other major railway companies were then adopting). In June 1870, as a witness before a House of Commons Select Committee on Railways, William O'Brien had remarked that the block system was 'premature' and no more than 'experimental'; public pronouncements that were to place him in direct confrontation with the Railway Inspectorate at the Board of Trade (BoT). Despite his objections, the Select Committee came to the conclusion that the adoption of the system of block working had "materially conduced to the safety of the public".

St. Nicholas' Crossing, a railway crossing on the level less than half a mile to the south of Carlisle station, was removed when the Midland Railway extended its route via Settle into Carlisle. Prior to this complete rearrangement of the lines, a very serious collision took place at that location on the 10th July 1870 when a West Coast night express from Scotland was cut into at right angles by a NER mineral train. The NER driver was under the influence of alcohol and either disregarded or did not observe the stop board. Five passengers were killed on the spot and many others were injured. On receiving news of the accident in London, George P. Neele, the London & North Western Railway (L&NWR) Superintendent of the Line, took the night express to Carlisle. William Cawkwell, the General Manager of the L&NWR, and George Leeman, the Deputy Chairman of the NER, accompanied him. They met at Euston station and, in discussion, Leeman impressed upon Neele

---

18 Apart from some sections of line in tunnels and one short stretch of the old Stockton & Darlington, the NER had chosen not to adopt the block system.

the desirability of avoiding any conflict between the companies at the official enquiry. Neele was subsequently to record that "unfortunately Captain O'Brien, his General Manager, was by no means so conciliatory when Col. Hutchinson, the Government Inspector, took evidence".[19]

By this stage the friction between William O'Brien and the Railway Inspectors was becoming obvious, and things were to get worse. Late in 1870, four serious accidents occurred on the NER, the last three of them within the space of eight days, and the inspecting officer who reported on them all to the BoT was Col. William Yolland. The first occurred at Scotswood Bridge, outside Newcastle, on 10th November; a collision that would not have occurred, so Col. Yolland considered, had the lines at that point been worked on the block system. Next came an accident, at Cargo Fleet, east of Middlesborough, on 30th November, which Yolland attributed in part to the variety of rules inherited from the constituent companies of the NER, obliging drivers on that particular section to carry two rule books and adjust their practices as they travelled out of one division of the company's system and into another.

A week later there was a disastrous collision between trains meeting head-on that resulted in five deaths. It occurred on 6th December at Brockley Whins, adjacent to the junction where lines from Newcastle, South Shields and Sunderland converged. Long overdue improvements to the layout and signalling had not begun; the points and signals controlling the junctions were not even interlocked to prevent conflicting train movements. In his report Col. Yolland came to the conclusion that: "Nothing can more plainly exhibit the entire absence of responsibility that exists on the part of the railway directors, their officers and servants, for the occurrence of preventable disastrous accidents, than what has taken place with reference to this very serious collision, because it appears to me that the company's management is wholly to blame for this accident." The last accident in this sequence followed just two days later, at Seaton, north of Thirsk. Once again Col. Yolland demonstrated that it would not have occurred if the block system had been in force.

The reports on these four accidents, in which Col. Yolland "tore into the directors of the NER"[20], were sent by the BoT to the company in the usual manner. In his report on the Seaton accident of 8th December 1870 Col. Yolland even took the opportunity to refer to some of the public comments made by Capt. O'Brien in the previous June. The longest of the reports — that on the Brockley Whins collision — was sent to the NER on 11th February 1871. The events on the NER had convinced Col. Yolland to try and persuade the BoT to obtain statutory powers to compel railway companies to adopt certain safety measures. The Chief Inspecting Officer, Capt. Henry W. Tyler, adopted a more cautious approach, proposing to a Select Committee that the practice of holding inquiries into railway accidents

---
19 G.P. Neele, *Railway Reminiscences*, 1904
20 Stanley Hall, *Railway Detectives*, 1990

CHAPTER TWO **Railway Relations**

should be legislated for. He considered that if this was done the BoT would get all that it required. Capt. Tyler's suggestion was incorporated into the Regulation of Railways Act 1871, and became an important milestone for the holding of inquiries into railway accidents.

Col. Yolland's reports were powerful public statements of dangerous weaknesses that the NER Board had to address. They recognised that the reports represented a general display of slack management, in several different quarters at once. William O'Brien had worked under one chairman only, Henry Stephen Thompson, and there was no evidence of any disharmony until the inquiry into the St. Nicholas Crossing accident. A quiet man by nature, he had displayed talents as a diplomat, discharging his duties as general manager apparently without arousing any opposition or complaint. He was of quite good standing in his profession and there was never any question regarding his probity. Against this, he does not seem to have been responsible for any major improvements in the manner in which the NER conducted its traffic working. As was the case with a number of other senior railway officers of the time, technical advances affecting the whole management of the railway had evidently gone by unnoticed.

The NER Board was so much disturbed by the disclosures that Capt. William O'Brien was asked to step down, and exactly a fortnight after the publication of the Brockley Whins report he left the service of the NER, his departure on 25th February not even being noted in Board minutes. There is no doubt that William O'Brien had been pressurised into resigning. If he had he been in ill health the Board would not have found it difficult to persuade him to retire and, under those circumstances, make a generous settlement as they did in the case of Thomas Cabry[21]. The last recorded mention of William O'Brien in the Board minutes was on 30th December 1870, ironically in connection with a report (to which he was a signatory) that recommended the installation of the block system on two sections of the NER mainline.

At the time of his retirement, William O'Brien had been General Manager of the NER for more than sixteen years. He must have been well liked as 5,494 members of the staff subscribed a total of £5,500 for a presentation to him, which was made on 14th September 1871. Following his departure from the NER, William O'Brien went to live at 23 Gloucester Gardens, London, within walking distance of his elder brother, Donatus. Whether or not the events of his last year on the NER had affected his health is not known, but he was to enjoy only eighteen months in retirement before dying at his London residence on 6th September 1872, leaving a

---

21 Thomas Cabry was a divisional engineer on the NER and resided close to Capt. William O'Brien at 31 The Mount in York. He retired at the end of June 1871 due to advancing years and failing health.

personal fortune of £45,000.[22] Subsequent to William's death, Margaretta moved to the Clifton district of Bristol, and later to Gloucestershire where she resided until her death at Barton Regis in 1890.

Returning to the Dromoland O'Briens, Sir Lucius Henry became the 3rd Baronet on the death of his father in 1765. Three years later he married Anne French, the only daughter of Robert French of *Monivea Castle*, Co. Galway. Their eldest son, Edward, born in 1773, acceded to the title in 1795 and four years later the 4th Baronet married Charlotte Smith, daughter of William Smith of *Cahermoyle*. They had thirteen children; amongst whom were Lucius (1800-72), William Smith (1803-64) and Henry (1813-95). Their first child, Lucius, who was born in December 1800, was the heir to the Dromoland estate and title, and on the death of Sir Edward in 1837 he became the 5th Baronet. At that time the Dromoland estate covered 20,321 acres. When the senior Inchiquin line became extinct in 1855 upon the death of James O'Brien, the 7th Earl, the Barony devolved to Sir Lucius who then became 13th Baron Inchiquin.

Sir Lucius was another member of the family to have an involvement with railways. When the Limerick & Ennis Railway (L&ER) was incorporated in August 1853 he was elected a director and, in the following year, he took over the chairmanship from Sir David Roche. A most interesting discovery was made during the construction of the line at Ballykilty, near *Dromoland*, in March 1854. Workmen found, concealed beneath a pile of stones, a small chamber containing a collection of beautiful gold ornaments, armlets, bracelets and collars dating from pre-Christian times. At that time it constituted the largest find of Bronze Age gold ornaments in Western Europe, thirteen of which are today in the National Museum collection in Dublin. Sir Lucius acquired four of the bracelets, one of which he presented to his second wife, Louisa Finucane, upon their marriage on 25th October 1854. Of the other three bracelets, Sir Lucius gave one each to Juliana, Charlotte Anne and Mary Grace, the three daughters by his first wife who were married, on the occasion of their respective weddings.

The L&ER was also interesting in that three members of the Maunsell family were involved with its Board: Dr Henry Maunsell, Henry Maunsell Esq. JP, of Limerick, and the company secretary Edward William Maunsell. Henry Maunsell Esq. JP, was a son of Daniel Maunsell Esq. and brother of George Meares Maunsell and Robert Maunsell, the latter being the grandfather of Richard Edward Lloyd

---

22 Simmons, in *The Express Train and other Railway Studies*, states that "this was a quite unusually large sum at that time for a man who, as far as we can tell, inherited nothing but made the whole of it by what he had earned as the salaried servant of railway companies". That comment was probably made in the absence of a detailed knowledge of Capt. William O'Brien's family background.

Maunsell.[23] Edward William Maunsell, who was also secretary of the Limerick & Castleconnell Railway (L&CR), was to leave the L&ER in 1862 to become secretary of the Dublin Wicklow & Wexford Railway, a position he held until retirement in February 1894.

Sir Edward and Charlotte's second son, William Smith O'Brien, born at *Dromoland* on 3rd October 1803, was educated at Harrow and at Cambridge. In the 1820s he took his seat in parliament as the Conservative member for Ennis and in 1835 he became the Conservative MP for Co. Limerick. In 1832 he married Lucy Caroline Gabbett, the daughter of Joseph Gabbett Esq. of *High Park*, Limerick. Joseph Gabbet's mother was the daughter of Revd Richard Lloyd of *Castle Lloyd* and his wife was Lucy Maunsell, the daughter of Ven. William Thomas Maunsell, archdeacon of Kildare. It was thus through William Smith's marriage to Lucy Gabbett that a further family connection was established with the lineage of Richard Maunsell.

Having gained parliamentary experience, William Smith's political views changed to a Nationalist perspective and he became a supporter of Catholic emancipation. He was a leading member of the Young Ireland movement and active in seeking relief from the hardships caused by the potato famine. He launched the Irish Confederation at Dublin in January 1847 but, pre-occupied with matters pertaining to the Grand Jury (predecessor to the County Council) for Co. Limerick, was unable to attend the first anniversary meeting of that body. However, he was persuaded to travel to Dublin for a three-day meeting to be held at the beginning of February 1848. On his way to Dublin by train from Ballybrophy on 29th January he penned some lines in Latin, an approximate translation of which was made by the late R.N. Clements:

> Now the railway carries us with a happy motion;
> *And heat drives the engine with clouds of steam;*
> *But the unpleasant smell makes one of us who is called Major sick;*
> *Nevertheless, we proceed with a vigorous turning of the wheels.*
> *There is no delay! We pass through the hometown of the Virgin;*
> *Then we reach the ancient shrine of St. Brigid;*
> *Now the high-tower of Clondalkin rises up;*
> *And the clear marbles of Dublin come into view;*
> *And soon we see its houses and churches*
> *"The end is the crown of any work", said the wise man; and so for us also.*
> *Our journey is finished —-we have finished our poem.*[24]

---

23 R.E.L. Maunsell commenced his railway career in 1888, as a pupil of H.A. Ivatt at Inchicore Works on the Great Southern & Western Railway (GS&WR). He later became Locomotive Engineer of that company and in December 1913 was appointed Chief Mechanical Engineer on the South Eastern & Chatham Railway. Upon the formation of the Southern Railway in 1923 he was appointed to the same position on the enlarged concern, a post that he was to hold until his retirement in 1937.

24 'The hometown of the Virgin' refers to Maryborough, and 'ancient shrine of St. Brigid' to Kildare. William Smith O'Brien's travelling companion, Mr Major, was Assistant Barrister for Co. Clare.

In 1848 William Smith urged the establishment of a National Guard, and an armed uprising was planned, but the famine had left the country without the spirit for such ventures. However, he did lead a small group of men who clashed with 46 policemen in Widow McCormack's cabbage garden at Ballingarry, Co. Tipperary, at the end of July. At 8.0 p.m. on Saturday 5th August 1848 at the recently opened Thurles railway station[25], William Smith attempted to buy a ticket. Billy Hulme, a former L&NWR employee who had been engaged by the GS&WR as a guard in 1847, identified him and seized him by the collar.[26] Detectives rushed forward, while a large body of policemen restrained William Smith 'in the roughest manner'.

Arrest of William Smith O'Brien at Thurles station on 5th August 1848 - [Author's collection]

Gen. John MacDonald interviewed William Smith at the Thurles Bridewell, offering him a bottle of champagne before proceeding to the station where he ordered the clerk, Arthur O'Leary, to provide a special train to Dublin. O'Leary demurred, and the General threatened him with a pistol, but the clerk explained that arranging a special train entailed a great deal of difficulty. Then, when the train was provided, Mulvany, the engine driver, declined to take any orders from the General until directed to do so by his superiors. The special train supplied to Gen. MacDonald's order consisted of a locomotive, two carriages, and a brake van, but unfortunately the locomotive number was not recorded anywhere.

William Smith O'Brien was sent for trial at Clonmel, the jury finding him guilty of high treason. Despite their disapproval of William's actions, the Dromoland O'Briens and friends rallied around. His brothers, Sir Lucius, Robert and Revd Henry, and sisters Grace, Anne and Harriet, all attended the trial. Robert and Revd Henry would attend court each day; and afterwards they would have interviews with

---

25 The railway station at Thurles, on the GS&WR mainline to Cork, was opened on 13th March 1848.

26 Billy Hulme received a £500 reward from the British Government, but did not live long to enjoy it. He resigned from the GS&WR on the Monday following William Smith's arrest, took a train to Dublin, and disappeared. Aged 28, he died in March 1850 at White Chimneys, Horton, Staffordshire, from the effects of hard drink.

counsel, talks with William and discussions with attorneys and lawyers. Following the guilty verdict, the death sentence was passed on William Smith, but this was commuted to transportation for life. Sir Lucius and Revd Henry endeavoured to procure his release by means of a petition, but they were not successful. William was imprisoned at Richmond gaol in Dublin and, before he boarded the naval brig *Swift* on 29th July 1849 for transportation to Tasmania, the whole family went to see him, taking part in a service conducted by Revd Henry.

After nearly five years in exile an unsolicited pardon was accorded to William Smith O'Brien on condition that he did not return to Ireland, and he then settled in Brussels. In May 1856 he was fully pardoned, and he returned to Ireland in July of the same year to reside at *Cahermoyle* once again with his family, but he avoided any further involvement in politics. All the newspapers recorded the remarkable co-incidence that occurred on his return journey home by the 4.0 p.m. train from Kingsbridge to Limerick Junction on 10th July 1856; it was hauled by the very same locomotive that had been used for the special train that had taken him to Dublin following his arrest in 1848. William Smith O'Brien died on 16th June 1864 whilst on a visit to Bangor in North Wales.

Sir Edward Donough O'Brien, 14th Baron Inchiquin, who acceded to the title in 1872 on the death of his father, Sir Lucius, was another member of the family who had a connection with railways. On 26th May 1884 an Order in Council was granted under the Light Railways Act of 1883 for the construction of the 3-ft. gauge West Clare Railway. Of the authorised capital, £100,000 was supported by baronial guarantee, the Barony of Inchiquin being one of the four guarantors. In recognition of the role played by the 14th Baron, one of the three 0-6-2 tank locomotives built for the company by Dübs & Co., Glasgow, and delivered in March 1892, was named *Lady Inchiquin*. The Lady in question was Sir Edward's second wife, Hon. Ellen Harriet White[27], to whom he was married on 29th January 1874 in London.

*West Clare Railway 0-6-2T No.7 **Lady Inchiquin** built by Dübs & Co in 1892 - [© IRRS collection]*

---

27 Ellen Harriet White was the eldest daughter of Sir Luke White, 2nd Baron Annaly, of Luttrelstown Castle, Co. Dublin.

Dromoland to Killiney

Henry O'Brien, Sir Edward and Charlotte's ninth child and fifth son, was born at *Dromoland House* on 15th April 1813. When he reached the age of 17 he followed in the family tradition and went to Cambridge where he studied law. Having graduated with a BA in 1835 he toured the continent, visiting Italy, Switzerland and France, and it may have been during these travels that Henry began to give serious consideration to a calling in the ordained ministry of the church. Whilst travelling from Como to Lake Lugano he was accompanied by a Swiss monk, and was later entertained by him and his Capuchin Brothers at their monastery where he engaged with them in some robust discussions on their respective views of the Christian faith. Henry travelled on through the Rhone valley, which he found exceedingly beautiful, and crossed the Simplon Pass on foot. In the south of France he went to hear a preacher by the name of Malan talk about the protestant churches in the region, but having met with him decided that he was not to his liking as a cleric. When Henry returned to Ireland he read theology at Trinity College, Dublin, and was awarded an MA in 1837.

The Church of Ireland parish church at Killegar, Co. Leitrim, where Revd Henry O'Brien was the perpetual curate, 1837-59 - [Author]

On 16th October 1837 the Revd Henry O'Brien was licensed as the district and perpetual curate[28] of Killegar, Co. Leitrim[29], in the diocese of Kilmore. The patrons of Killegar were the Godley family who had inherited estates in the townlands of Killegar and Drumergoul through marriage with the Morgan family. The church, together with its small adjoining glebe house, were built with an endowment of £1,100 from John Godley (1775-1863) a young Dublin merchant for whom *Killegar House* was built. *Killegar House* was completed in time for his marriage in April 1813 to Katherine Daly, who was a sister of 1st Baron Dunsandle. It was in that parish that the young curate soon became acquainted with Harriet, the elder daughter of his patron John Godley. John Robert Godley, Harriet's brother, was the founder of the province of Canterbuy, New Zealand, and his son, Sir John Arthur Godley, who was raised to the peerage as 1st Baron Kilbracken on 8th December 1909, was at one

---

28 Perpetual Curate — a priest nominated by a patron and licensed by a bishop to serve in a parish that did not have a vicar. Once appointed, such a curate had a lifelong tenure.

29 Some sources incorrectly show Revd Henry O'Brien appointed as PC Dowra in 1837. This error probably arose from the fact that the parish of Killinagh (Dowra) was the next entry to Killegar in published lists of the parishes in the united Diocese of Kilmore Elphin & Ardagh.

time private secretary to William Gladstone and later Secretary of State for India.

Henry and Harriet's relationship blossomed and they were married on 23rd May 1839. Three months later Henry conducted the service in Dublin when his brother Edward married Louisa Massy-Dawson. Henry and Harriet had seven children; two sons and five daughters. Their elder son, Edward O'Brien, who was born on 26th July 1840, joined the Indian Civil Service in 1863, and on 11th December 1866, at Hisar, Haryana, India, he was married to Mary Oclanis Lamb, daughter of Col. William Burges Lamb. By 1877 Edward was a Settlement Officer, 2nd grade, stationed at Bannu and three years later promotion to 1st grade also brought a move to Muzaffargarh. His report on the land revenue settlement of the Muzaffargarh District of the Punjab, 1873-80, was published in 1882. Edward's next move took him to Delhi as a Deputy Commissioner 2nd grade where, in 1887, he was the Acting Commissioner. He was promoted to Deputy Commissioner, 1st grade, at Kangra in 1890, and whilst there he wrote a paper on dialects of Kangra district. Edward O'Brien died at Dharmsala, Punjab, India on 28th November 1893.

Henry and Harriet's younger son, Murrough John, was born on 10th June 1842, but before reviewing the events in his life we shall take a brief look at the daughters in the family. Katherine, the eldest daughter born in 1843, was married to Revd Frederick Bransby Toulmin at Killesherdoney on 16th May 1867 and she died on 16th August 1912. Annabella Charlotte who, on 14th September 1871, married John Watt Smyth, a Judge of Lahore, died on 20th February 1907. Olivia Henrietta, the third daughter, was only 14 years old when she died on 21st May 1863, and the youngest daughter, Grace Amy Frances (b.1860), was married to Major Reginald L'Estrange McKerrell at St., George's, Hanover Square, on 26th May 1889, but sadly died nineteen months later on 28th December 1890.

On 4th June 1873, Henry and Harriet's fourth daughter, Angelina Rose Geraldine (b.1853), was married to John Gerald Wilson, C.B., of *Cliffe Hall*, Piercebridge, near Darlington. His grandfather, John Wilson Esq., who married Martha Bassett in 1799, purchased the Cliffe estate in 1825.[30] John Wilson died in 1836 but Martha lived on until February 1868. John's son, Richard Bassett Wilson (1806-67) was responsible for the rebuilding of the hall in 1859. Prior to that, on 5th December 1839, Richard Basset had married Anne Fitzgerald, sister of Mary the wife of Sir Lucius O'Brien (5th Bart Dromoland and later 13th Baron Inchiquin) and co-heiress of William Fitzgerald of Adelphi, Corofin, Co. Clare. Mary and Sir Lucius' third child, Charlotte Anne was actually born at *Cliffe Hall* in 1840, two years before the birth of John Gerald Wilson.

John Gerald and Angelina Rose's first child, Richard Bassett, was born in 1874 and

---

30 The Cliffe Hall estate lies on the south bank of the Tees, near the point where Watling Street passes straight across the river from Yorkshire into Co. Durham.

a second son, Murrough John (named after his mother's younger brother) followed on 14th September 1875. Murrough John Wilson was destined for a notable railway career. Educated at Marlborough College, he entered the service of the North Eastern Railway at 18 years of age in 1893 where he spent about nine years in the operating and locomotive running departments. However, he resigned his position in 1902 following a series of tragedies that resulted in him inheriting the Cliffe estate. Col. John Gerald Wilson, his son Lieut. Richard Basset Wilson and his brother Lieut-Col. Richard Bassett Wilson, all fell in the Boer War. Lieut. Richard Basset Wilson was wounded on 21st July 1900 at Oliphant's Nek whilst serving with the 3rd Battalion Imperial Yeomanry and died from his wounds five days later. Lieut-Col. Richard Bassett Wilson, CMG, while in command of 3rd Battalion Durham Light Infantry, died of enteric fever at Kroonstad on 21st March 1901, and Col. John Gerald Wilson was with the 3rd York & Lancaster Regiment when on 7th March 1902 he was wounded near Tweedbosch and died the next day. Angelina Rose Geraldine survived her husband by many years, it being 18th August 1941 before she passed away. We shall return to Murrough John Wilson's subsequent railway career in a later chapter.

Colonel John Gerald Wilson, father of Murrough John Wilson - [Author's collection]

In December 1859 the Revd Henry was appointed vicar of Denn, a parish 5¼ miles from Cavan town on the road to Ballyjamesduff, but it is doubtful if he was instituted to that parish as following the death of the Archdeacon of Ossory he was appointed to the parish of St. Mark's, Killesherdoney, where he was instituted on 23rd January 1860 as vicar. Killesherdoney is situated just 2½ miles to the south west of Cootehill on the road to Cavan in the townland of Cordoagh. In the 1800s there were several corn mills and whinstone quarries in the parish as well as a flax mill, but the Irish Mining Company's lead mine proved unprofitable and had been discontinued. The tragic death of Olivia Henrietta at Cordoagh Rectory in May 1863 has already been referred to; she was buried in the graveyard at Killesherdoney church on 21st May 1863 following a service conducted by Henry's brother-in-law, Revd James Godley.

St. Mark's, Killesherdoney, Cootehill, Co. Cavan. Hon & Revd Henry O'Brien incumbent 1860-75 - [Author]

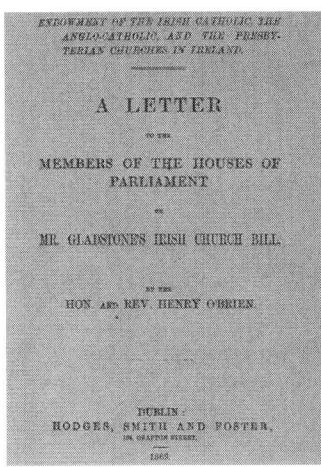

Cover of the Hon & Revd Henry O'Brien's letter on Gladstone's Irish Church Bill - [Author's collection]

In 1862 the surviving brothers and sisters of the 5th Baronet, with the exception of William Smith, had by Royal licence dated 12th September been given the style and precedence of the sons and daughters of a Baron. Therefore, it was the Hon. & Revd Henry O'Brien who was to write a long letter under the heading *Endowment of the Irish Catholic, the Anglo Catholic, and the Presbyterian churches in Ireland* to the members of the Houses of Parliament in 1869 in relation to Gladstone's Irish Church Bill.[31] A year earlier, when the number of rural deaneries in the united diocese of Kilmore, Elphin & Ardagh had been increased from eleven to thirteen by the addition of Bailieborough and Killesherdoney, Henry O'Brien was appointed Rural Dean of the latter.

Harriett died at 45 Fitzwilliam Square, Dublin, on the night of 30th April / 1st May 1872 and, following his resignation of the incumbency of Killersherdoney at the end of March 1875, Hon. & Revd Henry O'Brien sold his land (30 acres and 3½ perches adjoining the Glebe house), cattle, sheep, farming implements, carriages and horses, and moved to Windham, Glenhall, Great Glen, Leicester. After twenty years in retirement he passed away at Uxbridge, East Sussex, on 12th February 1895.

Murrough John O'Brien, Revd Henry's younger son, was educated at St Columba's College, Rathfarnham, Co. Dublin (1851-53) and at Radley (1855-59) where he rowed in the Radley eight. Following two years at Trinity College, Dublin, he worked his passage out to New Zealand aboard the sailing ship *Matoaka*. In New Zealand he spent time on the sheep stations in the Malvern Hills and followed the 'rush' to the gold diggings at Otago. The diggings became hopeless and so he walked the 150 miles to Dunedin and then took a steamer to Christchurch. In 1865 he was asked to assist Arthur Dudley Dobson in looking for a pass to the West Coast (now *Arthur's Pass*), but although Dobson got through he was murdered during his return. Murrough next managed Dalethorpe station (25,000 sheep) about 50 miles from Christchurch, but when that was sold, and the purchase of a larger station fell through, he left for Sydney. Having found no work up country he concluded that he should go home and, after two weeks in Sydney, he took a steamer to Melbourne where he spent a month. He sailed from Port Phillips on 4th July 1866 on board the *Reigate*, once again working his passage. It was very cold and stormy around the

---

31 The enactment of this Bill led to the disestablishment of the Church of Ireland

Horn, but he eventually arrived at Gravesend on 19th October having been 107 days out at sea.

In 1867 Murrough O'Brien went to Kenmare, Co. Kerry, as a pupil to William Steuart Trench. His agency, which was managed by his son John Townsend Trench, looked after Lord Lansdowne's estate.[32] Murrough became acquainted with the Mahonys of Dromore, who were very kind to him, and on his second visit to their house he fell in love 'at sight' with Mrs Mahony's sister, Ellen Waller. Born at *Shannon Grove*, Ellen was the third daughter of John Waller BL and Mary Franks, granddaughter of Bolton Waller (1769-1854), and a grand niece of the late Revd William Waller (1795-1863), Rector of Kilcornan. The parish of Kilcornan lies beside the river Shannon in the barony of Kenry, on the lower road from Limerick to Askeaton. The patrons of Kilcornan were the Wallers of *Castletown*, Co. Limerick, the parish church having been built in 1832 with a gift of £700 from John Waller. Castletown, the elegant residence of the Waller family, was situated in a richly wooded demesne of 200 acres that sloped gently down to the river. However, it was a long courtship and it was to be 16th September 1873 before Murrough and Ellen were married in Consul Graham's office at Bayonne. Their only child, Henry Eoghan O'Brien, who was in the 30th generation in line of succession to Brian Boru; a great-grandson of Sir Edward O'Brien, 4th Baronet Dromoland; a grand-nephew of Sir Lucius O'Brien, 13th Baron Inchiquin; and first cousin of Murrough John Wilson, was born on 24th August 1876. It was this child of nobility who was to become one of the 20th century's earliest proponents of mainline railway electrification.

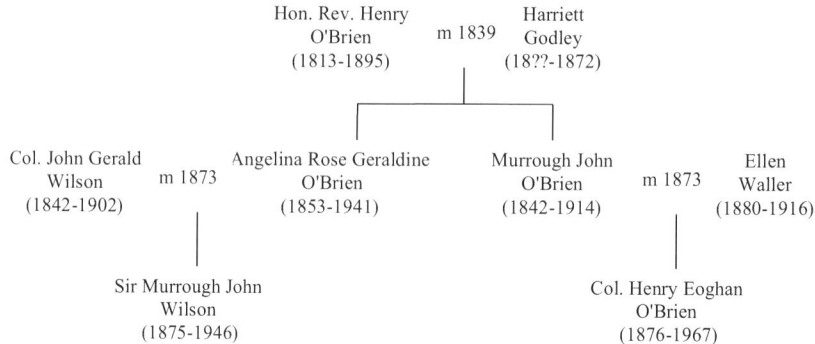

Family tree showing the relationship between H.E. O'Brien and Sir Murrough Wilson

In the intervening years, Murrough O'Brien had left the Kenmare office in 1869 to take a farm near Clones, Co. Cavan, but he was only there a short time when he got

---

32 Branches of the same Trench family were later to produce two notable railway engineers; Louis Trench (1846-1940), Chief Engineer of the Great Western Railway (1891-92); and Ernest Frederic Crosbie Trench (1869-1960), Chief Engineer of the North London Railway (1903-06), London & North Western Railway (1909-22), and of the London Midland & Scottish Railway (1923-30).

an offer to go to America to report on a plantation on Crooked River in Georgia. He found the place to be quite inaccessible, unhealthy, and the country disturbed with the Ku Klux Klan in power. Having got permission to leave America, he visited Niagara Falls on his way home and arrived at Queenstown (now Cobh) on 1st January 1871, the very same day that Gladstone's Irish Church Act of 1869 came into force. From that date the Church of Ireland ceased to be the established church; all rights of patronage were taken away, the crown no longer appointed the Irish bishops, all of the ecclesiastical corporations were dissolved, and its ecclesiastical law assumed merely contractual force. What attracted most controversy, however, was the issue of church property. In 1869, the Church of Ireland was the largest landlord in Ireland; some 11,000 tenants on 900 different estates throughout the island occupied its lands.

On disestablishment, the whole property of the church was vested in a lay corporation established under the Church Act — the Commissioners of Church Temporalities in Ireland. They were the successors to the Ecclesiastical Commissioners and their head office was located at 24, Upper Merrion Street, Dublin. In March 1871 Murrough John O'Brien was appointed Inspector of Estates and assistant to William Prendergast in the office of the Commissioners, and his work involved travel all over Ireland pricing church lands for sale. Following their marriage, Murrough and Ellen first lived in Dublin with Ellen's mother, Mary Waller, at 5 Ely Place Upper. In 1875 Murrough and Ellen moved to the southern suburbs of Dublin to reside at 1, Willow Park Terrace, Williamstown, and it was there, within sight and sound of DW&WR trains running along the line between Kingstown and the city that Henry Eoghan O'Brien was born, and where he spent the first years of his life.[33]

The Irish Land Commission, constituted as a rent-fixing agency under the Irish Land Act 1881, had wide-ranging judicial and administrative powers. Later it developed into a tenant-purchasing commission to facilitate the purchase and transfer of land from landlord to tenant and over time it was responsible for the transfer of some 13.5 million acres. The Land Commission was located at 24/25 Upper Merrion Street, and took over the responsibilities previously granted to the Board of Works under the terms of the Irish Land Act, 1870. It was also successor to, and stood in place of the Commission of Church Temporalities, but churches, burial grounds, schoolhouses and other church property were vested in the Representative Church Body (RCB)[34] by order of the Commissioners. The solicitor to the RCB was none other than John Maunsell, JP, the father of R.E.L. Maunsell, and there is no doubt that, as well as being related through marriage, John Maunsell and Murrough O'Brien would also have shared considerable contact in professional matters. Under such circumstances it is unlikely that Eoghan O'Brien and R.E.L. Maunsell

---
33 Eoghan is generally accepted to be the same as the Welsh Owen or the Scottish Iain, Celtic versions of John (Latin Johannes).
34 The Representative Church Body (RCB) was established by Royal Charter on 15th October 1870.

were unknown to each other; a factor that may have had a bearing on Eoghan's subsequent choice of career.

Murrough John O'Brien, Eoghan's father, on his appointment as Land Commissioner in 1889 - [Courtesy Prof. O.F. Robinson]

Ellen O'Brien (née Waller) Eoghan's mother - [Courtesy Prof. O.F. Robinson]

No.1 Willow Park Terrace, Williamstown, where Henry Eoghan O'Brien was born on 24th August 1876

Murrough O'Brien had been elected an Associate of the Institution of Civil Engineers of Ireland on 11th March 1878, and his role as Inspector of Estates was confirmed in the Sales Department of the Land Commission on 26th August 1881; his cousin Denis Godley being the Secretary to the Commissioners. Murrough took a great interest in land reform and wrote a good deal about it for *The Freeman's Journal*, the *Pall Mall Gazette* and the *New York Nation*.

It was just two years before the establishment of the Land Commission that Murrough purchased *Mount Mapas*, a modest house in Killiney that had been built about 1840. The county Dublin area of Killiney is noted for its hills commanding excellent views of Dublin Bay, the Vale of Shanganagh and surrounding mountains. It extends over a large area stretching from the bay across to Ballybrack and west to Glenageary, while to the north it includes Killiney Hill itself and is bounded by Dalkey. On the summit of Killiney Hill stands a pyramid-shaped obelisk, which was completed in June 1742 as the centrepiece of a grand project to plant and landscape the hill, which was at the time called Obelisk Hill. The obelisk is a distinctive landmark, and has been part of the Killiney scenery since it was erected by Col. John Mapas as a building project to provide some employment during the severe winter of 1741-42, which was a time of much poverty and suffering among the poor.

*The original single-platform Killiney station, as it appeared when Murrough O'Brien moved to **Mount Mapas** in 1879. The present Killiney & Ballybrack station, ¼ mile further south, was opened on 8th May 1882 following the doubling of the line from Dalkey to Ballybrack. - [Green Studio collection, courtesy Sean Kennedy]*

It was this special scenic quality which in the 19th century first attracted developers to build houses in this area, giving it the distinctive character which remains unchanged to this day as one of Dublin's most favoured residential quarters. Robert Warren, who owned the Killiney Hill estate for much of the 19th century, repaired the obelisk in 1840. He also facilitated the Dublin & Wicklow Railway (D&WR) by allowing the line between Bray and Dalkey, which that company built in 1853-54, to traverse his property at the Vico fields. In addition to compensation paid to him by the Railway Company, he was also provided with a halt, known as Obelisk Hill, and a footbridge that gave access to White Rock. Although Obelisk Hill was initially a private halt when the line opened on 10th July 1854, the D&WR agreed in June 1855 to open it up for public use on a temporary basis. However, it did not last long and it was decided to build a new station at Killiney, ⅝ mile further south, once a roadway had been provided. In September 1857 Robert Warren wrote to the D&WR stating that the road would be ready on 1st October, and requesting that the new station should be commenced. Obelisk Hill station was closed on and from 1st January 1858; the new Killiney station being opened on the same day.

The original line from Bray to Dalkey was single track with a junction at Shanganagh for the inland route to the city via Dundrum, which at that time was also a single-track line. However, the section between Shanganagh Junction and Bray was soon doubled, the additional track being brought into use on 30th October 1855. Doubling of the line from Dundrum and remodelling of the junction was completed by 15th July 1861 by the Dublin Wicklow & Wexford Railway, to which the D&WR had changed its name under an Act of 15th May 1860 that also authorised an extension from Wicklow to Enniscorthy. The physical junction at Shanganagh was dispensed with in May 1877, three separate tracks being provided into Bray — two for the Harcourt Street line and a single one for the coastal line from Dalkey. From the time that the coastal line was opened there was also a station at Ballybrack, 1¼ miles south of Obelisk Hill. On 8th May 1882, following the doubling of the line between Ballybrack and Dalkey, a new combined Killiney & Ballybrack station replaced the separate ones at a point halfway between the two. The 1882 station is in use to this day for Dublin Area Rapid Transit (DART) commuter services and the two older station houses can still be identified.

After the opening of the railway, there was a rush to build houses in Killiney with the dual advantage of magnificent views and convenience of being within easy reach of Dublin city. Most of the residential development took place in an area extending south from Killiney village towards Ballybrack and the sea, but some properties, like *Mount Eagle*, had been built before the arrival of the railway. These large Victorian houses were built to reflect the status and position of the families that lived in them. Some of the earliest 19th century houses in Killiney were villas, and *Mount Mapas*, located near *Mount Eagle*, is a good example of this type. It stands one-storey over-basement and its first floor ironwork veranda endows it with a particularly colonial air. *Mount Mapas* appears to have been vacant from 1867 until 1872 when it was then occupied by William Colles, Surgeon, whose consulting rooms were located at 21 St. Stephen's Green, just three doors away from where John Maunsell conducted his legal practice. William Colles lived at *Mount Mapas* until his death, sometime after 1878, when his son, Graves Chamney Colles, sold the property for £2,000 to Murrough John O'Brien who moved in with his family on 6th November 1879. Having settled in Killiney and secured his position with the Land Commission, Murrough took his family on a holiday to Switzerland in 1884, a country that he had first visited in 1869, and thereafter holidays in Switzerland became an annual event.

Mount Eagle, a Killiney property in a remarkable location with unparalleled views overlooking the sea, had, when completed in 1837, originally been named *Coburg Lodge* in recognition of the Saxe-Coburg family name of the new monarch, Queen

Victoria. For a time it was later also known as *Kent Lodge* before assuming its better-known name. It was designed by Sandham Symes and constructed completely in squared and jointed blocks of ashlar granite stone, the symmetrical 'Y-plan' being a unique feature of the house. The design in the neo-classical style, has an attractive double staircase, and is decorated inside with plasterwork in the Greek Revival style. The six-acre site features a broad terrace that looks out over the Irish Sea, and the stables, which back onto the road, have cut stone details.

Sandham Symes was a nephew of Robert Warren and was commissioned by him to design many new buildings around Killiney Hill. A fine wooden model of *Mount Eagle*, which used to be displayed in the house, bore Symes' signature.[35] Among other buildings in Killiney designed by Symes was Holy Trinity (Church of Ireland) church. Robert Warren gave the site for this church and also made a substantial donation towards the cost of its construction. It was completed in 1858, and the first incumbent was Revd Joseph Samuel Bell (1831-91) who served the parish until 1867.

Mount Eagle, Killiney, Co. Dublin. The unique symmetrical 'Y-Plan' of the design is evident in this view - [M.J. O'Brien; courtesy Prof. O.F. Robinson]

The 'Eagle' on Killiney Hill which overlooked *Mount Eagle* - [G.F. Woodworth]

Mount Eagle was vacant for some years prior to 1873, when Sir Ralph Smith Cusack, Chairman of the Midland Great Western Railway, occupied it. He moved to Furry Park, Raheny, in 1875, and by 1878 Graves Chamney Colles was in residence at *Mount Eagle*, and lived there until his death in November 1892. Murrough O'Brien acquired the property from Mrs Colles in June 1895 for £3,500, including furniture, but he continued to reside at *Mount Mapas* and let *Mount Eagle* to Wilfrid Fitzgerald until September. It was then taken for a very short period by John Morley, the Chief Secretary for Ireland (1892-95) who had just lost his seat at

---

35 A nephew of Sandham Symes, also Sandham, was appointed head of the Audit Department of the GS&WR in January 1872 and retired on 1st November 1906. In turn, his son, Sandham John Symes, also joined the GS&WR as a pupil of Robert Coey at Inchicore. S.J. Symes left the GS&WR in 1903 to gain further experience with the North British Locomotive Co at their Atlas Works in Glasgow. He joined the staff of the Midland Railway in January 1904 being promoted to Chief Draughtsman at Derby in July 1913. In October 1925 he became Works Manager at Derby and in April 1928 he was appointed personal assistant to Sir Henry Fowler.

the 1895 election.[36]

Murrough John O'Brien had first made the acquaintance of John Morley in October 1880 when he sent him an article for *Fortnightly Review*, and was on friendly terms with him right up to the time that he became Chief Secretary. Morley was well aware of Murrough's views on land reform and taxation and that he was often the recipient of hostile criticism. It was on the family's return from holiday in Switzerland in the summer of 1892 that they learnt of the death of John George McCarthy, one of the Land Commissioners. Several friends wrote to Murrough suggesting that he should apply for the post, but he would not go to the Chief Secretary's Lodge in the Phoenix Park to talk to John Morley about the position.

However, Ellen went to the Lodge to call on Mrs Morley and took Eoghan along with her, as he was interested. Mrs Morley was very nice to them and talked a good deal to Eoghan, but when Lady Wolseley came in they left the Lodge and Eoghan went to see the Zoo. Mrs Morley returned the visit soon after, and then in October 1892 Murrough and Ellen were invited to lunch at the Chief Secretary's Lodge, and at the end of the month Murrough was appointed a Land Commissioner at a salary of £2,000 p.a. Murrough and Ellen moved into *Mount Eagle* on 10th October 1895, and having vacated *Mount Mapas*, it was let to Rt.Hon. C.T. Reddington who died there on 4th February 1899 after a month's illness. In December 1899 *Mount Mapas* was let for five years to W. Patterson, manager of the *Daily Irish Independent* newspaper.

Murrough O'Brien was a noted figure in the city of Dublin, always to be seen dressed in Donegal tweeds. He held a deep interest in Irish archaeology and also studied the Irish language. Among personal friends were two notable supporters of Irish Home Rule, Michael Davitt and John Dillon. Of his appointment as Land Commissioner it was said that it would have been absolutely impossible to select anyone more likely to discharge the duties of the office with efficiency, prudence and equity; "to great practical experience he also brought a cultural and liberal mind". Morley said of him "A more scrupulously conscientious upright man does not exist in Ireland today", and in a personal letter to Ellen he confided: "It is not often that a man is so lucky as to be able to show his appreciation of a private friend while recognising a faithful public servant. I am delighted at all that has happened." Murrough worked hard to achieve fairness in land sales, but the landed gentry received every proposal that he made with hostility and suspicion. He served as a Land Commissioner until the passing of the Wyndham Land Act in 1903, when land purchase activities were transferred to the Estates Commissioners, and he was left with no work of importance. Murrough O'Brien retired on 1st August 1904 after

---

36 Lord Morley was well noted for his integrity and sympathetic approach to the question of Irish Home Rule. He assisted Gladstone in carrying the second Home Rule Bill (1893) and attacked privileges of the Anglo-Irish Ascendancy, removing from office some of the more intransigent magistrates.

33½ years service and received a pension of £1,333 6s 8d per annum.

From *Mount Eagle* it was a short distance down Vico Road and along Station Road to *Court-na-farraga*, the residence of William Exham, Esq., who lived there from the time it was completed until his death about 1881. Situated just past Killiney railway station, it was specifically designed for him in 1865, probably by the architect T.N. Deane, in a fanciful French chateau style with a series of ornamental stone-capped gables, and a conical tower, which were all elaborately surrounded by a two-storey veranda. Following William Exham's death, his widow continued to live at *Court-na-farraga* until 1887, but by 1888 the property was in the hands of John Burke, Esq., who remained there until 1894. After that it appears that it was vacant for a number of years until purchased in 1899 by Capt. Richard Altamont Smythe, at which time Francis B. Ormsby, Secretary of the GS&WR, who resided at *St. Anne's*, would have been a nearby neighbour. In the 1970s, after *Court-na-farraga* ceased to be a private residence, it was extended and became the well-known Killiney Court Hotel.

Court-na-farraga, Killiney, Co. Dublin, the residence of the Smythe family from 1899. - [Courtesy Prof. O.F. Robinson]

Richard Altamont Smythe (1838-1924) had been a Captain of the Salop Militia and on 29th July 1869 he married Frances Anne Jane Bellingham, 4th daughter of Sir Alan Edward Bellingham (1800-89). Sir Alan was educated at Trinity College, Dublin, where he received a BA in 1821, and also attended Exeter College where he obtained an MA in 1832. On 26th August 1827 he became 3rd Baronet of the Bellinghams of Ireland. He was Deputy Lieutenant for Co. Louth and High Sheriff of the same county in 1829, a position that was also held by Richard Altamont Smythe in 1871. Richard was the eldest son of Henry Meade Smythe of Newtown, Co. Louth, and Frances Barbara Cooke, daughter of Rev Richard Cooke of Ballyneal, Co. Kilkenny. Prior to their move to Killiney the Smythe's had lived at *Lauragh*, one of the principal seats of residence in the parish of Coolbanagher, Queen's County (now Co. Laois). Their arrival at *Court-na-farraga* brought to the neighbourhood a bright 14-year old girl; Frances Victoria Lucy Smythe, who had been born at *Lauragh* on 3rd March 1885, and who would later catch the eye of her young neighbour from *Mount Eagle* — Henry Eoghan O'Brien.

The Formative Years

For the young Henry Eoghan O'Brien early life at *Mount Mapas* and *Mount Eagle*, overlooking the DW&WR line between Killiney and Dalkey, would have provided him with ample opportunity to observe railway activities. It is not known whether or not the sight, sound and smells of steam locomotives got 'into his blood' at an early age, but living in such close proximity to the railway must have had some influence on the young boy. An interest in mountain walking and climbing was developed at an early age, he being brought up to it by his father who was an early member of the (British) Alpine Club. Eoghan was at first taught at home by a Miss Hill and Mlle Laure Nicole of Lausanne, but when he was older he attended Leonard Strangways' school on St. Stephen's Green in Dublin. His academic progress may be judged from the following letter, which he penned in January 1887, at the age of 10, to the Editor of *Harper's Young People*:

> I am an Irish boy living on the coast opposite England, and I would like to let American boys and girls know the Irish side of the question. Ireland once had a Parliament of her own, but it was partly under the control of the British Government, they ordered that its members should only be Protestants. A great many of these were very corrupt and were bribed by the British Government to vote for the union of the two countries. It succeeded, and they were united. It was only five years ago that the franchise was so extended to allow the heads of families to vote. In America, I understand it is much more extended. Things were in a bad state fifty years ago, but they are better now. Twenty years ago there were some terrible evictions at a place called Glenveigh; they tore the roofs off the houses, they pulled old men and women out of the houses while their children stood crying by.

Mlle Laure Nicole of Lausanne with her sister at **Mount Eagle** *- [Courtesy Prof. O.F. Robinson]*

In January 1889 Ellen became very unhappy about Eoghan's schooling. He had to go in to town every morning by the 7.48 a.m. train and he did not arrive home until just 4.0 p.m. having had only a bit of bread and butter to eat. After partaking of some of his mother's tea, he would then go out for a stroll up Killiney Hill with his dog and then it was in to lessons again, often until 9.0 p.m. This had not been the case before Christmas 1888 as Mr Strangways had allowed him to come out by the 12.46 p.m. train and he had two hours to get home for lunch. An added pressure was the fact that he was working with boys older than himself who were studying for a scholarship, and he was often given work that was beyond him. Ellen happened to mention her concern in a letter to Murrough's sister Kate (Toulmin) who passed the message to Geraldine (Wilson). Geraldine wrote back to Ellen suggesting that Eoghan might be sent to Mr H.E. Hill, who had been tutor to her boys for some years and had prepared Dick (Richard Bassett) Wilson for Eton. Mr Hill lived in a cottage three minutes walk from *Cliffe Hall* and the boys went to him every day for their lessons. And so it was agreed, Eoghan setting off from Killiney with his father on 8th February on his way to *Cliffe Hall*.

Geraldine constantly wrote to say how well Eoghan was doing and Mr Hill said that he worked very well. Eoghan's name had been put down for Charterhouse, and in June 1889 Ellen travelled to London and met up with Eoghan's grandfather, Revd Henry O'Brien, and they went on down to Charterhouse together. Although she found the new school buildings very nice and picturesque, she did not take to Dr Haigh-Brown, the headmaster, finding him ponderous. She even had to ask him six times before she got to see Mr Locke, the housemaster, who seemed a nice person and gave them tea. Ellen left Charterhouse unsettled in her mind as to what would be best to do about Eoghan's education when they returned from their annual visit to Switzerland.

Some years earlier Ellen had met and made friends with E.L. Vaughan (1851-1940), a housemaster at Eton College, so when in London she arranged to go down to Eton and meet with him. She told him of her concerns about Eoghan's future, but being aware that a boy's name had to be down for a long time for entry to Eton she felt that it would be impossible for him to go there. Vaughan informed her that it so happened a vacancy had arisen and he would be able to take Eoghan, so Ellen wrote to Murrough to tell him of the situation. While the O'Briens were on holiday in Switzerland, a letter arrived from E.L. Vaughan informing them that he had to have a prompt decision as he had just received another application. Murrough made up his mind and, having agreed that Eoghan could go to Eton, he telegraphed 'accepted' to Vaughan.

Ellen was delighted, as she knew that Edward Littleton Vaughan, being a highly conscientious and active young man, would do all that a housemaster could. Eoghan passed his entrance examination and entered Eton in September 1889 where he was placed in the middle fourth. On 24th October Vaughan wrote of his initial impression of the new boy: "O'Brien has been well taught and is clear-headed. He is far in front of the average boy of his age and promises well. On the other hand, as compared with the picked boys with whom he is classed, he does not come much to the front. He is not as fast or as accurate a worker as several others and could not expect to be very high in the class in an examination. He is in everyway a good boy and is likely to get on well. He ought to continue to work in the 'first set' with the best boys and seems to have ability and industry enough to do so. I fancy that he has done before nearly all the work he has done with me so far, but he has not had practice in it so as to acquire facility." His mathematics master added, "I should say that for his age he is very well on and about equally so in the three subjects; arithmetic, algebra and Euclid".

When Eoghan came home for Christmas 1889, Murrough went to Kingstown pier to meet him and was surprised when a fellow he saw in a tall hat and a long greatcoat turned out to be his son! A year later Eoghan was in fifth form, had joined the Army class, gained distinctions, and was generally progressing very well. By May 1892 he had come second in the junior school for the Prince Consort's prize for French and in early 1893 he was rowing in one of the Eton eights. In the second heat for the summer term boat race he had to stand in at No.2 thwart in Thorpe's crew at very short notice and received special mention in the *Eton Chronicle* for his 'plucky rowing'. He made a trip to London with E.L. Vaughan to see the Mint and St. Paul's Cathedral, and they ended their visit in the House of Commons. There he met Mr Asquith, who showed little interest until Vaughan introduced him as 'son of Mr Murrough O'Brien of the Land Commission' at which Asquith seemed much more enthusiastic and shook Eoghan's hand warmly. On 12th July 1893 he was up in London again, spending the whole day at Lords for the Eton *v* Harrow cricket match, which Eton won.

A view from Eton College; a photograph taken by Eoghan in 1894. - [H.E. O'Brien; courtesy Prof. O.F. Robinson]

Eoghan O'Brien (on the right) with some of his peers at Eton College in 1895. - [Courtesy Prof. O.F. Robinson]

In September 1893 Eoghan was made captain of his House at Eton and this gave him a good deal of responsibility. It was originally anticipated that he would go to the Royal Military Academy at Woolwich to study engineering, but as the time for the entrance examination drew near his parents thought that his mathematics might not be good enough to get him into the Royal Engineers, and they felt that the Royal Artillery was an expensive and useless profession for Eoghan as he had no vocation to be a soldier. However, he did sit the Woolwich examination in November, although his masters at Eton did not expect him to succeed. While he was in London for the examination, his mother went over to be with him and she took the opportunity to see E.L Vaughan, who gave her a very good account of Eoghan. Although they did not accept any invitations to dine out, as Eoghan wanted to be quiet and get to bed early, on one occasion they did dine at the House of Commons where they saw a good many interesting people, including Lord Randolph Churchill.

When Eoghan came home to *Mount Mapas* at Christmas he was accompanied by a German governess, Fräulein Neumann. She was the governess at *Cliffe Hall*, and Geraldine Wilson had offered her to the O'Briens for a Christmas engagement to help Eoghan with his German lessons, and she taught him splendidly. In March 1894 Revd Henry O'Brien, his grandfather, was out with his aunt Kate (Toulmin) for a day and they met with Eoghan for lunch at the *White Hart* hotel in Windsor. In a letter to his mother Revd Henry wrote:

> "Eoghan arrived looking very nice and well in his white tie, but he looked a little tired, and with good reason, for he had had a very hard day on Thursday with the volunteers at Aldershot. Kate and I thought him charming, with quite gentlemanlike manners, at ease and very pleasant in telling us his manner of life, his studies and amusements. We went up to his room, which was comfortable, but not too luxurious as Eton boys' rooms are often said to be. The walls were covered with themes of home, and two photographs of his grandfather."

Eoghan's Easter vacation was foreshortened, as he had to go back early to take extra lessons with a mathematics tutor at Eton. However, he did enjoy his short break as his father had hired a typewriter, with which he was quite fascinated, and they both learned how to type. It was during the Swiss holiday of 1894 that a telegram arrived from E.L. Vaughan advising that Eoghan had in fact passed the examination for the Royal Military Academy, in which he came 17th out of the 50 who passed. On the family's return from Switzerland at the beginning of September there was a telegram waiting for them, which said that Eoghan had passed the Higher Certificate of the Oxford & Cambridge examination board in Mathematics, French, Latin, Physical Geography and Geology that he had sat after the Woolwich exam. He also gained the Geology & Physical Geography prize at Eton in the summer of 1894.

In October it was arranged that Mr Baker, the National School master at Eton, should come in to the college twice a week to give Eoghan lessons in bookkeeping. Then in November a great flood occurred and Eton was under water for ten days. Eoghan and his tutor were nearly drowned when their boat was upset in the flood and they were both hurled into the foaming torrent. He was involved in another serious incident when he was on his way home for Christmas on 20th December. The 1,431-ton RMS *Munster* was involved in a collision and had to put back into Holyhead, the passengers being quite shaken, and when Eoghan arrived home he was still quite pale.

However, recovery was quick, as a few days before his arrival a telegram had come to say that Eoghan had gained 1st place in the Prince Consort's Prize for French, a great honour and an award of books and £15. In sending congratulations to his grandson on his success, Revd Henry wrote: "The £15 will I have no doubt be useful, but the chief reward is that it tells that you have worked well and accurately and have obtained a thorough knowledge of the French language." During the holiday there were occasional dinner parties, which he enjoyed, but he did not care to go to the many dances, preferring instead the charms of the fireside at *Mount Mapas*. As an adventure away from home, Murrough took Eoghan on a short tour of Belfast and up to the Giant's Causeway where he observed the famous tramway – the first hydro-electric railway in the world. It is not known if that visit had any influence on the decision to put the name *Brian Boroimhe* on a new locomotive delivered to the tramway in 1896. Murrough was also busy trying to settle Eoghan's future and through Sir David Dale, a close friend of the Wilsons, he ascertained that Sir James Kitson would take him into the Airedale Foundry at Leeds to begin mechanical engineering when he finished at Eton.[37]

We have already noted the death of Eoghan's grandfather on 12th February 1895 in Chapter Three. Murrough went over to Uxbridge, and together with Gerald Wilson and Eoghan followed the small funeral to the grave. Murrough had to go over to England again in mid-March to be questioned at Westminster by the Financial Commission as to the financial relations between England and Ireland. When that was over he went on to Leeds to look out rooms for Eoghan and also had a pleasant meeting with Sir James Kitson who suggested that Eoghan might call on his daughters! As Easter approached, E.L. Vaughan wrote pleading with Eoghan's parents to leave him at Eton for just one more term, even offering to take him 'free gratis and for nothing'. He had also met Sir James Kitson at the House of Commons and had ascertained that it did not matter to Sir James when Eoghan started. It was subsequently learnt that a start in August would be too late, and so he commenced his practical engineering training at the Airedale Foundry in April 1895. After Eoghan had left Eton, one of his masters wrote of him:

> "He made a good captain and I could nearly always go to him when I was in trouble about noise, etc., besides which I found him a real kind, helpful friend and he cheered me up, for even if he only had time to pop his dear kind face in at the door, he had something bright to say. He was truly liked and there was a general void when the sad news came that he was no longer an Eton boy and one of us. He was so good and brave about it all — I can tell you better of him than I can write."

So, what were the factors that might have influenced a pupil from Eton to choose Leeds as the place to further his education? In 1837 James Kitson established an engineering workshop in Hunslet, a suburb of Leeds immediately south of the river Aire, and about two years later these works were known as the Airedale Foundry. The

---
37 A niece of Sir David Dale's wife, Sybil May Millbank, was later to marry Murrough John Wilson.

firm soon developed a reputation for engineering excellence and the construction of quality locomotives — winning a medal at the Paris International Exhibition of 1867. Engineering training with Kitsons was much sought after by many young men in Victorian times and during the 1880s two sons of another Irish nobleman, the 3rd Earl of Rosse, entered the firm. One of them, Richard Clare Parsons, who was a partner in the firm from 1880 to 1887, interested himself keenly in steam tram engines; some 300 examples of which were built by the firm from 1879/80 onwards. It was Richard who was instrumental in attracting to Kitsons' premises his brother, Hon. Charles Parsons, the future inventor of the steam turbine.

During the 1880s Charles Parsons set himself the task of achieving very high-speed rotation of a shaft, but he had only moderate steam pressure at his disposal. In a shed at the Airedale Foundry he worked out designs for a high-speed engine of an original character and entered into an arrangement with the partners of the firm (his brother being one of them) that they should carry on work in an alliance based on special terms. However, the in-service performance of the novel features of his 'epicycloidal machine' led Charles Parsons to conclude that this invention was not the solution, and he turned his mind to another concept – the steam turbine. He also invented and made torpedoes at the Airedale Foundry, but these were a secondary matter in the Parsons-Kitson alliance. Although the design for the steam turbine was not to be developed in his experimental shed at Kitsons, Charles Parsons never forgot his time with the firm. On the very day in 1927 when he became the first engineer upon whom the Order of Merit was bestowed, Charles Parsons returned to the Airedale Foundry to show his former colleagues the medal. He later wrote that he attributed to Leeds "a realisation of the difficulties of high speed reciprocating machinery".

Ellen O'Brien on her bicycle in 1894 - [M.J. O'Brien; courtesy O.F. Robinson]

The aforementioned Paris International Exhibition of 1867 was the motivating force behind the establishment of the Yorkshire College. Visitors from the West Riding of Yorkshire, particularly from the woollen and textile industries, had been alarmed at evidence of the rapid development of new technologies on the Continent, which appeared to them to pose a considerable threat to the local cloth trade. They were particularly concerned at the superior scientific and technical education of European workers. On their return from Paris, James Kitson and Obadiah Nussey (a woollen manufacturer) proposed to the Yorkshire Board of Education the establishment of 'a county college of science'. They succeeded in obtaining support of other industrialists and the London-based Clothworkers' Company offered generous backing.

In 1874, the teaching of mathematics, experimental physics, geology, mining and chemistry began at the College; and biology was soon included. After a few years, classics, modern literature and history were added and the Yorkshire College of Science became simply the Yorkshire College. The Yorkshire College combined with the Leeds School of Medicine in 1884 and three years later the two Leeds-based institutions joined Owens College, Manchester, and University College, Liverpool, as constituent members of the Victoria University. The new federal university had hardly got into its stride before each of the big cities started to consider the benefits of having their own independent universities. When Liverpool and Manchester decided to establish their own universities, independence was forced on Leeds and in 1904 King Edward VII granted the Leeds University its own Charter.

It was to this city of Leeds, with its ability to provide a fine combination of technical education and practical engineering training that Eoghan went in 1895. Exactly why he came to the decision to adopt a railway-engineering career is not known. Whether it was prompted by a love of railways generated from days spent watching the trains of the DW&WR pass along by Killiney, or by a suggestion from family connections in the industry, the end result was the same. At the time he started at the Airedale Foundry the firm was under the directorship of the late James Kitson's two sons; John Hawthorn Kitson (1843-99) and his brother James Jnr.; Thomas Purvis Reay was the Managing Director and the Chief Draughtsman was Thomas Haslehurst Brocklebank.

During his time at Kitsons Eoghan was to commence a lifelong friendship with fellow student, William Herbert Morton. Morton was appointed Assistant Works Manager at Kitsons in 1899, but in August 1900 he moved to Ireland as Chief Draughtsman with the Midland Great Western Railway (MGWR) at the Broadstone in Dublin. Morton was a Leeds man, through and through, having been born in Hunslet in 1877, and his move to Dublin is somewhat difficult to explain unless it was influenced through his friendship with O'Brien. Although there is no proof, it is not beyond the bounds of possibility that he could have been a holiday guest at Killiney. In such circumstances, the two young railway engineers may have taken the opportunity to visit the various railway facilities in Dublin, including the MGWR workshops at the Broadstone.

W.H. Morton was to have an outstanding railway career. In 1915 he was appointed Locomotive Engineer of the MGWR and, following the amalgamation of all railways operating wholly within the Irish Free State to form the Great Southern Railways (GSR) in 1925, he became deputy to John R. Bazin, the Chief Mechanical Engineer of the enlarged organisation. When Bazin retired in April 1929, Morton

CHAPTER FOUR **The Formative Years**

W.H. Morton, Locomotive Engineer MGWR (1915-24) - [Author's collection]

was the natural choice to succeed him. The pinnacle of his career was attained in April 1932 with his appointment as General Manager of the GSR; a position that he held until retirement at the end of September 1941. As we shall see in a later chapter, H.E. O'Brien was to frequently correspond with Morton on the question of electric traction on the DSE section of the GSR.

In April 1895, when Eoghan started his practical engineering training at Kitsons, the works was busy finishing Order No.152 for the Waterford & Limerick Railway (W&LR). That order was for two new 0-4-4 tank locomotives, which were basically an update of a design that had originated in 1894 with the rebuilding by J.G. Robinson of an 0-4-2 goods locomotive as 0-4-4 tank locomotive No.15 *Roxborough*. The W&LR placed the order with Kitsons on 2nd August 1894, and works numbers 3587 and 3588 were assigned. The first locomotive, No.51 *Castle Hackett*, was being steamed at Kitsons by May 1895, and Eoghan would have been thrilled to see No.52 completed, proudly bearing the name of his royal ancestor *Brian Boru*.

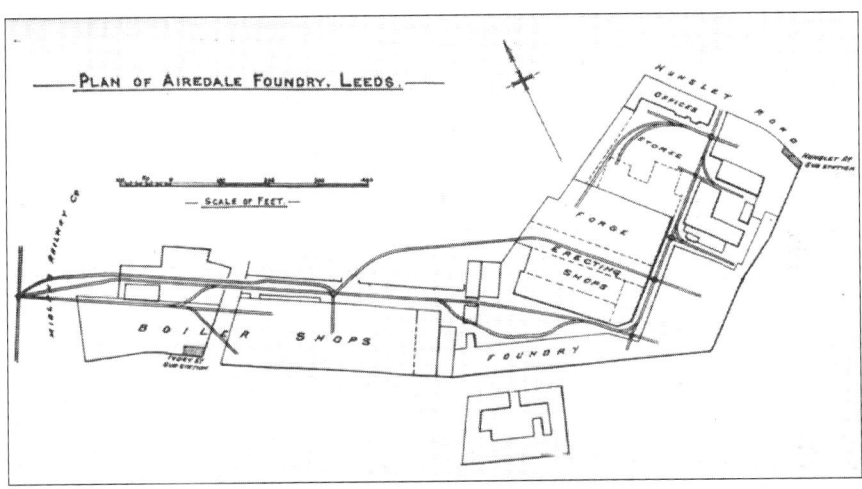

Plan of Messrs Kitsons' works, Airedale Foundry, Leeds - [Author's collection]

The workmanship must have been to Robinson's satisfaction, for just three months later the W&LR went to Kitsons again, this time for two 4-4-2 tanks (Order No.160) and two 4-4-0 tender locomotives (Order No.161). Works numbers 3616-19 were assigned and work had commenced on them before Eoghan departed from Kitsons to begin his studies at the Yorkshire College of the Victoria University in October 1895. The four locomotives were delivered between February and April 1896, by which time the company had changed its name to the Waterford, Limerick

& Western Railway (WL&WR).[38] Kitsons must have proved satisfactory once again as an additional two tank locomotives and one tender locomotive, to the same designs, along with three 0-6-0 goods locomotives were ordered from them on 21st May 1896.

*WL&WR 0-4-4T No.52 **Brian Boru**, built by Kitson & Co., Leeds, in 1895 photographed at Sligo shed on 13th September 1898. - [H.L. Hopwood, LCGB H-758 © NRM, York]*

When Eoghan returned to Killiney for Christmas 1895, the family had already moved in to *Mount Eagle* and he was quite taken with the new home. The library had been coloured a little like that at *Mount Mapas*, a new carpet had been fitted, and the books had all been arranged as it was to be his room. As Eoghan felt that there would be very little more to be gained by returning to Kitsons in the summer of 1896, he seized the opportunity during the holidays to go into Dublin and arrange with T.B. Grierson, the Engineer of the Dublin Wicklow & Wexford Railway at Westland Row, to work in the shops of his 'home railway' at Grand Canal Street during his college vacations.

Eoghan was admitted as a student member of the Institution of Civil Engineers on 19th May 1896, and he arrived home for his summer vacation on 23rd June. After a holiday in the Tyrol and Engadine with his father, he started at Grand Canal Street. Murrough would get up at 4.40 a.m. and make coffee and send him off, and Eoghan would arrive back about 6.0 p.m. 'black as a sweep'. As time went by, he had not only a pass for travelling by train, but also a locomotive pass, and he enjoyed riding on the footplate.

---

38 The Waterford & Limerick Railway, which had purchased the L&ER and L&CR on 1st January 1873, became the Waterford, Limerick & Western Railway (WL&WR) on 1st January 1896. The company's Locomotive Engineer from 1889 to 1900 was John George Robinson.

CHAPTER FOUR **The Formative Years**

At that time locomotive engineering matters on the DW&WR were in what might be kindly described as 'a state of flux'. As a result of adopting a report into the affairs of the company by W.C. Furnivall, M.I.C.E., which had emanated from an investigation that he had carried out in 1894, several senior officers, including the Locomotive Superintendent, William Wakefield, had resigned. The DW&WR Board had then placed all responsibility for locomotive, carriage and wagon matters in the hands of the Chief Engineer, Thomas Benjamin Grierson. Although Grierson was well versed in railway civil engineering matters, he lacked knowledge and experience of locomotive practice and design. This was to result in a poor specification for a class of 4-4-0 passenger locomotives ordered from the Vulcan Foundry in 1895, which turned out to be weaker performers than their smaller 2-4-0 predecessors, being bad steamers as well as suffering from broken frames.

The 1882 Killiney & Ballybrack station; **Mount Eagle** *is the last house visible in the distance on the right. - [Author's collection]*

Grierson's next venture was with goods locomotives for the heavily-graded New Ross branch and, having drawn up a specification, he suggested five firms from whom tenders could be obtained, viz. Kitsons, Neilsons, Beyer Peacock, Sharp Stewart, and the Vulcan Foundry. The order placed by the DW&WR with Kitsons in April 1896 for two 0-6-2 tank locomotives at £2,575 each was to result in what has become known as 'Grierson's Folly'. Although the locomotives on Kitsons Order No.1/4 were to the same basic design as those built by them for the Rhondda & Swansea Bay Railway and the Lancashire Derbyshire & East Coast Railway, they were unlike any DW&WR locomotive. Eoghan would have been aware of this for he would probably have known all of the DW&WR locomotives quite well by virtue of his own observations from the vantage point of his home at Killiney. The young pupil engineer returned to the Yorkshire College in October 1896, but it is not known if he called in on Kitsons to see these two locomotives under construction while he was at Leeds.

49

That all was not well in the Locomotive department of the DW&WR is evident from a resolution passed at Board following a series of serious locomotive failures: "in the opinion of the Board it is desirable to provide in future for the administration of the Locomotive & Carriage department separately from the Permanent Way department and, therefore, that immediate steps be taken to find a suitable person to act as Superintendent thereof".[39] The DW&WR was in the process of making enquiries for a suitable person to appoint as Locomotive Superintendent when Grierson reported on his inspection of the first 0-6-2 tank locomotive. He stated that he had found the tests satisfactory, and advised that the second locomotive would be ready for testing about a week later. Candidates for the post of Locomotive Superintendent were interviewed on 26th April 1897, and their testimonials were examined; four being selected to come before a special Board meeting three days later. After full and careful consideration by the Board it was decided to appoint Richard Cronin, who at that time was Principal Foreman at the Inchicore works of the GS&WR.

The arrival of Cronin at Grand Canal Street workshops signalled the beginning of a renaissance in the locomotive affairs of the DW&WR, but his first job was to deal with the problem of the new tank locomotives that he had inherited from Grierson. Unlike the 4ft 8½in. gauge versions, which weighed only 58t 4cwt, the two DW&WR examples were considerably heavier; most of the additional weight being on the trailing driving wheels. As a result of this they were prone to derailments in yards and sidings and, following one such incident on 4th June 1897 in the North Wall yard of the GS&WR, Cronin turned for help to Robert Coey, his former chief at Inchicore. He arranged to have both locomotives weighed and the findings on the distribution of weights on the axles as ascertained at Inchicore were reported to Board, as follows:

	Specification	Loco No.4	Loco No.5
Leading driving axle	15t 00cwt	16t 08cwt	15t 15cwt
Centre driving axle	15t 03cwt	16t 17cwt	16t 19cwt
Trailing driving axle	15t 05cwt	17t 03cwt	17t 00cwt
Radial axle	15t 05cwt	14t 10cwt	14t 06cwt
Total weight	**60t 13cwt**	**64t 08cwt**	**64t 00cwt**

Cronin reported the weights of No.5 on 10th June 1897, but there was a gap of three weeks before he submitted those of No.4. H.E. O'Brien's return to the DW&WR as Cronin's pupil, following the completion of his second year at the Yorkshire College, could well have been in time for him to see No.4 being weighed at Inchicore. If this was the case he would probably have met with his distant relative Richard Maunsell, who was the Works Manager there, and Cronin may also have

---

39 DW&WR Board Minute No.2797, 17th December 1896

taken the opportunity to introduce his pupil to Robert Coey, the Locomotive Engineer of the GS&WR.

R.E.L. Maunsell, Locomotive Engineer GS&WR (1911-13), Chief Mechanical Engineer SE&CR and SR (1913-37) - [Author's collection]

'Grierson's Folly' - DW&WR No.4, one of the two 0-6-2T locomotives purchased from Kitsons of Leeds in 1895 for the heavily graded New Ross line, photographed ca 1900 at Harcourt Street shed. - [H. Fayle collection © IRRS]

Strangely, it was Grierson who reported on Kitsons' letter of 8th July regarding the locomotives, but the Board referred the matter to Cronin for a report. Richard Cronin made his views known a week later when it was ordered that Kitsons be written to in terms of his report. Kitsons replied on 27th, but it was obviously an unsatisfactory answer for the DW&WR who, in turn, replied by repeating their request for Kitsons to send a man over to see the engines weighed. In late September Kitsons offered a reduction of £30 per locomotive, but the DW&WR rejected this offer and sought £75 off each in settlement. This was obviously agreed to, as the matter was not raised again. The settlement appears to have been a small amount for locomotives that were clearly not suited to the task for which they were purchased, and it was probably an indication of the DW&WR's acceptance of Grierson's part in the matter. Having the locomotives built for the wider Irish track gauge must have been a contributory factor to the increase in weight, a point that Grierson should have been conscious of.

Although not strictly related to Eoghan's career, it is worth noting subsequent events. T.B. Grierson resigned on 24th February 1898, having secured a position as Chief Engineer on the Lancashire, Derbyshire & East Coast Railway, but his last two locomotives were to survive somewhat longer with the DW&WR and its successors. In August 1904 Cronin submitted a proposal to alter them into tender locomotives, at a cost of about £600 each, so as to reduce the weight on any pair of driving wheels to not more than 15 tons; tenders being ordered from the Vulcan Foundry in January 1905 at £558 each. New boilers with Belpaire fireboxes were obtained in 1920 from the Yorkshire Engine Company and fitted in 1924-25; the locomotives becoming Nos.448 and 449, works class J1, on the GSR in 1925. The

GSR withdrew No.449 in 1940, but No.448 lasted into CIE days before it was taken out of service in 1950.[40]

It would have been an interesting time for Eoghan on the DW&WR in 1897, as he watched his new chief get to grips with matters at Grand Canal Street. He would have been there while the design for the first of a new series of 2-4-2 suburban tank locomotives was developed. The DW&WR was in urgent need of the new suburban tanks in order to effectively respond to increased competition from the Dublin United Tramways Company. The vastly improved tramway service that had come about as a result of introducing electric traction on the Dublin-Dalkey route from 16th May 1896 was having a disastrous effect on the DW&WR receipts. Eoghan O'Brien could not help but be aware of the situation, and that the DW&WR was even considering electrification of the Amiens Street to Kingstown (now Dun-Laoghaire) section as a means of improving its suburban train service. This particular problem on his 'home railway' could well have sown the seeds of his own interest in electric traction.

Richard Cronin's first design: DW&WR 2-4-2 suburban tank locomotives, No.3 ***St. Patrick*** *-*
[Author's collection]

Grand Canal Street works would have provided H.E. O'Brien with his first exposure to the construction of new passenger rolling stock; two six-wheel 28ft 0in. first class carriages with four-compartments, Nos.47 & 48, being built there in 1896. He would also have observed the construction of passenger brake vans and horseboxes and the reconstruction of many goods wagons. Small railways like the DW&WR were of insufficient size to warrant separate running departments, and working under the Locomotive Superintendent would have presented an opportunity for Eoghan to learn about railway operating practice. An accident that occurred on the system at Wicklow Junction during Christmas vacation work with the DW&WR,

---

40 The DW&WR changed its name to the Dublin & South Eastern Railway (D&SER) from 1st January 1907, and became a constituent of the Great Southern Railways (GSR) on 1st January 1925. The GSR and the Dublin United Transport Company amalgamated on 1st January 1945 to form Córas Iompair Éireann (CIE).

would hardly have escaped his notice either, and is worth describing.

DW&WR First Class saloon carriage built in 1895 at Grand Canal Street works. - [K.A. Murray collection © IRRS]

In the early hours of the morning of Friday 3rd December 1897 the DW&WR ballast train was loaded with gravel in the Chemical Works yard at Wicklow Junction. At 4.15 a.m. the locomotive propelled the train out of the yard and onto the 'up' mainline and, having pulled forward clear of the Wicklow Junction crossover and uncoupled from the train, it proceeded to Wicklow station to cross over to the 'down' line so as to run back around the train to bring it to Dublin. The points of No.8 crossover at the Dublin end of Wicklow station, which were normally used for this movement, were not set for the driver to cross, so he proceeded on the 'up' line as far as the Wicklow station signal box. On enquiring why he was not let cross, the signalman informed the driver that he did not know the locomotive was coming.

The driver then proceeded to the points at the south end of the station where the signalman let him back onto the 'down' line, showing him a green light from the signal box. Noting that the points were correctly set, the driver proceeded back along the 'down' line towards his train and, having passed by the 'down' platform, whistled for a signal from the Wicklow Junction signal box. Seeing a green light, the driver put on steam and ran round the curve towards his train, not realising until he was about 30 yards from it that the station signalman had, inexplicably, reversed No.8 crossover and turned him back onto the 'up' line through the points at the Dublin end of the station.[41] Although he reversed his engine and made every effort to stop, he could not do so before colliding with the ballast train. The brake van and two wagons were damaged and two men who were in the brake van, Ganger

---

41 The distance between the No.8 crossover, at the Dublin end of Wicklow station, and the crossover at Wicklow Junction was approximately 640 yards on a continuous left hand curve of 18 chains radius.

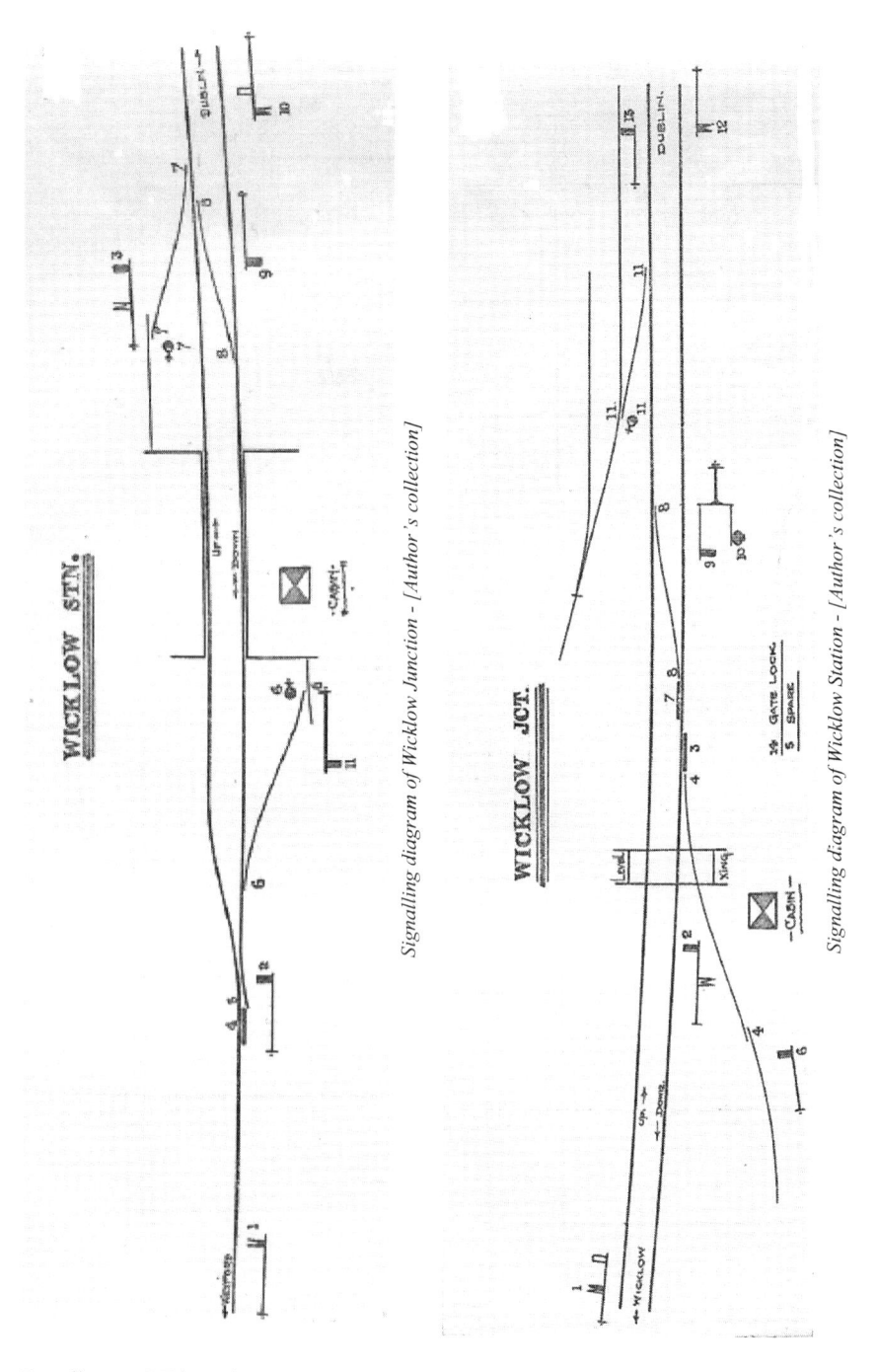

Signalling diagram of Wicklow Junction - [Author's collection]

Signalling diagram of Wicklow Station - [Author's collection]

Dowling and Edward Hogan, were injured. A.M. Ford, an inspecting officer from the BoT, held an inquiry at Wicklow on 18th December 1897, and he found the station signalman at fault. The signalman was reduced to the grade of platform porter and the ballast train guard was reprimanded for not adequately supervising the shunting operations.

Compared to the Airedale Foundry, the DW&WR provided Eoghan with much wider exposure to railway engineering and operating matters, but by Christmas 1897 it was becoming evident that it was never going to be a big enough railway for him to gain the kind of in-depth experience that he needed for his chosen career. John Morley had enquired of Murrough if he could help in any way and wrote to Yates Thompson, a Director of the Lancashire & Yorkshire Railway (L&YR). Richard Cronin recognised that his pupil had considerable potential, and knew that he also had to do something to assist in the young man's progression. From his Inchicore days Cronin was well acquainted with J.A.F. Aspinall, Chief Mechanical Engineer of the L&YR. Aspinall had been the GS&WR Locomotive Superintendent at Inchicore from 1882 to 1886, and it was no doubt through the Coey-Maunsell-Aspinall network of locomotive engineers that Eoghan's next move was finally forged in 1898. It is quite possible that Richard Maunsell had spoken highly of his own experience at Horwich in 1891-92 as an Aspinall pupil, and that such a testimonial from his relative convinced Eoghan to go there after he had completed his engineering studies at the Yorkshire College of the Victoria University.

Eoghan went over to Manchester on 8th January 1898 for an interview the next day with Aspinall, and it was arranged that he should go to Horwich in the following September as one of his pupils. For his last few months at Leeds he moved out of the city to live at *Briary Farm*, Bramhope, and he cycled the six miles between the farm and the Yorkshire College on a regular basis. Shortly after returning to Killiney for the summer, a telegram was received on 30th July from John Goodman, Professor of Engineering at the Yorkshire College, offering 'hearty congratulations' to Eoghan on his obtaining a 1st Class B.Sc. degree from the Victoria University. In a letter that followed Goodman added, "May it be but the first fruits of a long and successful career — honest hard work always reaps its reward, not only at 'compound', but at 'quadruple expansion' interest". During the early summer of 1898, Eoghan had also sat the examination required for admission as an Associate Member of the I.C.E. and on 26th November he was notified that he had been successful in that regard.

On 5th September 1898, two bright young men of noble descent walked through the gates of the L&YR Works at Horwich. Herbert Nigel Gresley and Henry Eoghan O'Brien were about to commence their training as pupils of J.A.F. Aspinall. William Barton Wright, Locomotive Superintendent of the L&YR, had resigned on 23rd June 1886 and Aspinall, at that time Locomotive Superintendent of the GS&WR, was appointed as his successor on 14th July. He took up the position on 1st October 1886 with the new title of Chief Mechanical Engineer (CME). The land at Horwich upon which the L&YR built their workshops had been purchased at auction on 27th May 1884. Prior to that, on 18th October 1883, John Ramsbottom, formerly

Locomotive Superintendent of the L&NWR, had been appointed by the L&YR as Consulting Engineer in matters relating to the Locomotive Department. Together with Barton Wright, he went over all aspects of the department and, in February 1884, he reported to the L&YR Board that the existing works at Miles Platting and Bury were inadequate and incapable of expansion to meet the company's needs.

Ramsbottom was largely responsible for the planning and layout of the new Horwich locomotive works in conjunction with Barton Wright. Work on the construction of the workshops commenced on 9th March 1885 and the official opening took place on 15th November 1886, just six weeks after Aspinall's arrival. Very soon after taking up his appointment he recognised the need to provide proper training facilities and courses for the staff at the newly established workshops. The L&YR Board sanctioned a grant of £2,500 to provide an Institute for the recreation and education of their workpeople at Horwich. Pending the erection of premises, Aspinall, James T. Tatlow[42] and others, started evening classes in the works offices in September 1887 in the subjects of steam, applied mechanics, machine drawing and mathematics.

The Horwich Mechanics' Institute - [HOR-E14 © NRM, York]

The new premises, which comprised two Class Rooms, a small Lecture Hall, Library, Reading Rooms and Smoking Rooms, were completed in early December 1888; the Horwich Railway Mechanics' Institute being formally opened on 15th of that month by George J. Armytage, Chairman of the L&YR. As the years went by, a great deal of difficulty was experienced in accommodating the influx of students, and the lack of the necessary classrooms prevented the establishment of new classes for which there was urgent need. The Directors of the L&YR responded by granting a further sum of £2,500, supplemented by a loan, and a large extension of the Institute was begun. The new portion was opened on 27th July 1893, and comprised two Class Rooms, a new Library, News Room, Magazine Room, Smoking Room, and a large Lecture / Concert Hall; the older portion thereafter being entirely devoted to technical school work.

Further extensions were the Fielden Gymnasium[43] opened in 1895, and the Chemical and Mechanical Laboratories, which were brought into use in 1901. The Recreation Ground, which was opened on 23rd April 1892, was developed from

---

42 James T. Tatlow, had been appointed accountant for the Locomotive, Carriage & Wagon department at Inchicore in 1881. He was encouraged to follow his old chief to Horwich in June 1887.

43 The construction of the Fielden Gymnasium was facilitated by a fund set up with an endowment of £1,500 from Sarah Jane Fielden, the widow of Samuel Fielden, late director of the L&YR. The first Trustees of the fund were John A.F. Aspinall, Henry A. Hoy and Philip Smith.

grazing land into a park, and included a cricket pitch and a bowling green, with pavilions. Other parts were afterwards used for playgrounds; a second bowling green (opened in 1895), and tennis courts, which were opened for play in May 1905, being later additions. Added attractiveness to the Recreation Ground was lent by a bandstand of handsome design, first used in June 1907. The L&YR also built a large dining hall to provide their staff with convenience for breakfast and dinner at the works. It could accommodate 1,100 men who brought their own food, which was cooked or warmed for them as required. The Café, built contiguous to the dining hall in 1889, was intended to meet the needs of those staff who wished to purchase meals. Although at first managed by a company from Manchester, its running was taken over in June 1902 by the Institute Committee.

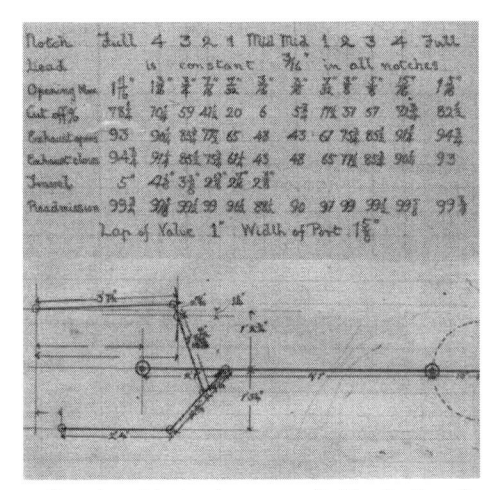

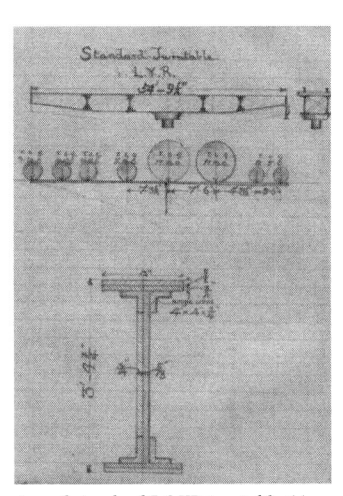

Drawing of Joy's valve gear and table of setting (l), and drawing of standard L&YR turntable (r) made by Eoghan O'Brien during his pupilship at Horwich, 1898-99. - [Courtesy Prof. O.F. Robinson]

At Horwich, in an environment designed to provide for all the needs of L&YR staff, young men developed their skills and knowledge of railway mechanical engineering. The President's Prize, for the student who made most progress in the first session of 1887-88, was presented to Henry Fowler. He followed up this success in 1891 when he gained a Whitworth Exhibition, the first to be won by a Horwich student and a distinction only shared with John Robert Billington and William Alfred Barnes. Other distinguished students in the earlier years of the Institute's work were Gervase Henry Roberts, Oliver Winder, John Peachy Crouch, Thomas Cross Hutchinson and George Nuttal Shawcross, and Reginald George McLaughlin, James Harold Haigh, William Arter and William Henry Marsh Parr deserve mention for success in later years.[44]

During the time that Nigel Gresley and Eoghan O'Brien were pupils at the Mechanics Institute their teachers for steam were Henry Fowler and John R. Billington. In the Science and Art Examinations for 1899, H.E. O'Brien was very successful in the subjects of Machine Drawing and Steam, and in the Commercial and Technical Examinations he qualified in Mechanical Engineering. He was also admitted as a graduate member of the Institution of Mechanical Engineers (I.Mech.E.) on 27th April 1899. In the following year he was successful in Machine Drawing, Theoretical Inorganic Chemistry, and Practical Inorganic Chemistry; and shared the Institute Prize for Practical Inorganic Chemistry with Charles E. Reed.

At a meeting of students of the Institution of Civil Engineers held on 23rd February 1899, with J.L. Thornycroft, FRS, in the chair, Eoghan presented a paper, jointly with Barnard Humphrey[45] a fellow student and friend from the Yorkshire College, on the subject of *Bearing Springs*. The paper addressed the use of springs and the conditions required for high efficiency. Pointing out that springs for similar work varied greatly in design, and that there was a great variation in the formulae used for deflections, etc., the authors noted that springs were unknown 300 years earlier, and that many improvements had been effected in the suspension of vehicles during the 19th Century. Recalling that the theory of leaf springs was first thoroughly investigated in 1852 by Edouard Phillips (1821-89), a French railway engineer, they referred to two methods of arriving at deflection and gave a résumé of Phillips' research work. The formulae arrived at by Daniel Kinnear Clark were also given, and the main points of a paper on railway springs presented by Sir Benjamin Baker at the I.C.E. in 1881 were noted.

Following a description of the process for manufacturing springs, the authors detailed their experiments with spring steel in tension as well as compression tests on both leaf and spiral springs. In those tests where the Young's Modulus of the spring material was known, the deflection calculated by Phillips' formula was found to agree closely with the practical results. The large variation in efficiency, on the basis of deflection per ton of load per pound weight of material in the

---

44 W.A. Barnes entered the service of the L&YR as an apprentice on 1st October 1894. He was appointed a draughtsman in 1899 and Chief Electrical Draughtsman in January 1915. Promoted to Assistant Resident Engineer, Clifton Junction in July 1915, he retired from that position with the LMS on 31st August 1926.
W.H.M. Parr left the L&YR in 1910 to join Harland & Wolff and was one of the nine-man 'guarantee group' who were lost in the sinking of the RMS *Titanic* on 15th April 1912.

45 Barnard Humphrey, born in 1875, was educated at Rossall College, Fleetwood (1886-92) and was another Kitson pupil (1894-97), also attending the Yorkshire College (1893-94). He remained with Kitsons after completing his apprenticeship in September 1897. In January 1899 he went back to the Yorkshire College to carry out experimental work under Prof. John Goodman and it was probably as a result of that activity that the joint paper with H.E. O'Brien was written. In September 1899 he joined the GWR at Swindon as a fitter and in October the following year he entered the drawing office. He was elected an Associate Member of the I.C.E. on 8th April 1902 and later took charge of the Motor Car department of the GWR in the office of the Superintendent of the Line at Slough.

CHAPTER FOUR **The Formative Years**

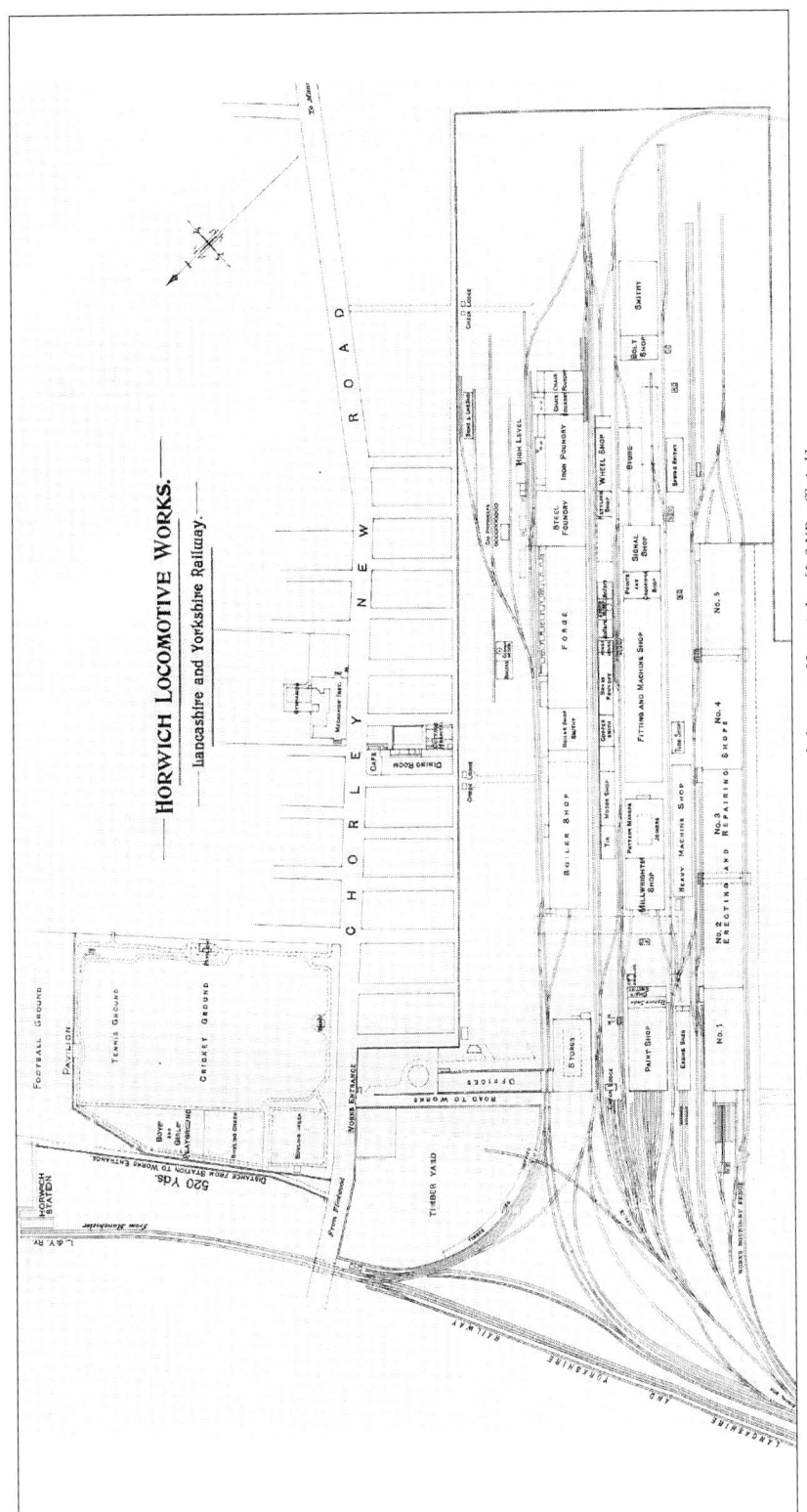

General plan of L&YR locomotive workshops at Horwich - [L&YR official]

spring, was noted. The authors compared spiral springs with leaf springs and the suitability of the former in certain applications was pointed out, and experiments showing the differences in resilience of a laminated leaf spring and a spiral spring were described. A discussion followed in which one of the participants was the young Charles Stewart Rolls.[46] The paper was considered to be the best student contribution in the session, and on 6th November 1899 the authors were awarded a Miller Prize by the Institution.

While he was a pupil, and until the summer of 1903, Eoghan lodged at 73 Victoria Road with Thomas Clayton, his wife Alice and their three children, John, Stanley and Doris. At that time Clayton was a slotting machinist at Horwich Works but he was later to rise to the position of Foreman of the L&YR Electric Motor Shop. Eoghan became close friends with Nigel Gresley during their time as pupils, and they were often invited to dine with the Aspinalls at *Fern Bank* where they also played tennis with the Aspinall daughters, Isabel and Edith. Although he had to put in long hours at Horwich, Eoghan still managed to get home to Killiney for Christmas and Easter, and it was at Easter 1899 that he first met the Smythe family. They had recently moved into *Court-na-farraga* and they dined with the O'Briens at *Mount Eagle* on 1st April.

When Eoghan O'Brien and Nigel Gresley started their time at Horwich, Aspinall's three senior assistants were the Works Manager at Horwich, Henry Albert Hoy, who had been appointed to the same post at the old works at Miles Platting in 1886 before his transfer to Horwich on 12th April 1887; George Hughes, who had been appointed Assistant Carriage & Wagon Superintendent on 23rd October 1895 following the retirement of Frederick Attock; and Charles O'Keeffe Mackay who was in charge of the Outdoor Locomotive department, ably assisted by George Banks. The Works Manager at Newton Heath was James Howarth and his assistant was Oliver Winder. Hughes' Outdoor Assistant in the Carriage & Wagon department was Arthur Dansey Jones, who had just succeeded James Davies on 1st August. This all changed with Aspinall's appointment as General Manager of the L&YR on 1st July 1899. Henry Hoy moved up to CME, George Hughes became Assistant CME and Works Manager, Horwich, and George Banks was made Assistant Carriage & Wagon Superintendent. A.D. Jones became Mackay's assistant in the Outdoor Locomotive department and G.H. Roberts filled the vacant position in the Outdoor Carriage & Wagon department. Then, on 1st October 1899, Oliver Winder took over from George Banks on the latter's appointment as Assistant Passenger Superintendent.

Eoghan's pupilship continued uninterrupted under Henry Hoy, and by February 1900 he was in the pattern shop, which was clean work, but by July he was in the heat and dirt of the forge. By Christmas 1900 he was in possession of a 1st Class

---
46 C.S. Rolls and Henry Royce went into partnership in 1906 to form the famous Rolls Royce company

CHAPTER FOUR **The Formative Years**

pass, which he used to make a return trip on the GS&WR Mail train between the pier at Kingstown (now Dun Laoghaire) and Queenstown. His return journey to Horwich was via Belfast, travelling from Kingstown pier in the through carriage that the Great Northern Railway (Ireland) attached to the Mail train, and in passing through Belfast he visited the Harland & Wolff works. When at home Eoghan would quite often meet the Smythes, and by Easter 1901 they had become dear friends to the O'Brien's. One Sunday night during a visit to *Mount Eagle* at the end of June 1901, when sitting out after his father had gone to bed, he discussed his future with his mother while she played the harp, and they sat together for a long time while watching the reflection of the moonlight on the waters of Killiney Bay. At that time the Aspinalls happened to be on holiday at Bray, and Murrough and Ellen called on them, but although Ellen only got to see Mrs Aspinall, Murrough was able to have a word with Mr Aspinall.

Horwich Works – (l to r) Paint Shop, Running Shed and Erecting Shop - [HOR-F-467 © NRM, York]

Liverpool – Southport Electrification

Following the completion of his time as a pupil, Eoghan O'Brien entered the service of the L&YR in the hydraulic department at Horwich on a salary of £100 per annum. He subsequently spent time on various experimental work and in the drawing office. C.O. Mackay died suddenly on 29th May 1901, and the consequential moves that took place on 1st July resulted in A.D. Jones taking charge of the Outdoor Locomotive department with J.P. Crouch as his assistant. At the same time H.N. Gresley, who had taken over from G.H. Roberts in the Outdoor Carriage & Wagon department when the latter moved to the Gas department at the end of June 1900, became Assistant Works Manager at Newton Heath in place of Crouch; Gresley being replaced by Charles Hubert Montgomery. It was around the same time that Aspinall instructed O'Brien to thoroughly familiarise himself with all aspects of electric traction.

During the period September to November 1901 Eoghan was in London investigating electric working on the Central London Railway. It was the first tube line to traverse central London and was, when opened on 30th July 1900, initially worked by electric locomotives. The annual task of the line equated to 40.442 million ton-miles, and the equivalent energy consumption was 0.102 BTU per ton-mile, or 13.55 BTU per train-mile. In gathering information, Eoghan also took note of a discussion on a paper by Edward Carlisle Boynton entitled *Notes on Electric Traction under Steam Railway condition*s, which was held at the American Institute of Electrical Engineers in March 1900. He also investigated the design and characteristics of three-phase generators and motors, and the relative advantages and disadvantages of employing alternating or direct current were examined.

Aspinall himself had benefited from a visit to America early in his career, and he was a great believer in exposing his staff to the experience of overseas railway operations as part of their development. He was also aware of the advantage that the L&YR could derive from such visits by their officers, particularly to America. He had sent George Hughes and Henry Hoy there in April 1899, and in February and March 1901 the party that went across the Atlantic comprised William Barton Worthington, Chief Engineer; Charles W. Bayley, Chief Traffic Manager; Samuel Hauxwell, Assistant Goods Manager; Oliver Winder, Assistant Carriage & Wagon Superintendent; and James T. Tatlow, Chief Clerk.

Following their return in September, Aspinall had formed a small committee comprising Hoy, Winder and O'Brien who were tasked with looking into all aspects of working the Liverpool, Crosby and Southport section by electric traction. Traffic on the Liverpool to Southport line had been growing on a year by year basis and, despite the provision of quadruple track on the section between Bank Hall and Seaforth during 1886-87; it was becoming ever more difficult to provide an adequate service. Various ideas had been considered, including an extension of the quadruple track beyond Seaforth, but the cost of purchasing land was prohibitive, as the area served by the line was a favoured residential district that was developing rapidly. At its meeting on 31st October 1901, the L&YR Board authorised Aspinall to spend up to £3,000 for the purpose of experimenting with electric traction.

H.E. O'Brien was elected an Associate Member of the Institution of Civil Engineers on 14th January 1902, and on 25th February he attended the annual dinner of the Manchester branch of the Institution at which he proposed the toast to 'our guests'. He was to recall that he did not make a good speech, but was able to go ahead smoothly for about five minutes and made his audience laugh once. He was, however, glad to do it as he felt it was very good practice. James Howarth, the Works Manager at Newton Heath, retired on 28th February 1902 and was replaced by H.N. Gresley with Francis Edward Gobey as his assistant, and on 27th May Eoghan met up again with Gresley when they dined together.

Towards the end of April he had spent fifteen days on a trip to Germany inspecting electric works with Henry Hoy and Oliver Winder. Their first visit consisted of two days at Schückertwerke Elektrizitäts AG in Nürnberg, where that company assigned an engineer to look after them and also placed an electric motor car at their disposal. Then, after a brief call to a die pressing works and the electric light power station in Frankfort, they spent time at Lahmeyer's works before proceeding on to Mainz, Wiesbaden, Cologne and Dusseldorf. On their return, Aspinall and the team entered into discussions with Dick, Kerr & Co. Ltd, Preston, and by July 1902 details of an electrification scheme were becoming clear enough for the Board to approve an additional £1,000 for experimental work. Before travelling to Switzerland to join his parents on their annual vacation in the summer of 1902, Eoghan was a guest at the wedding of Aspinall's daughter Edith to Smelter Joseph Young.

By 22nd October 1902 Aspinall was able to present his proposal to the Board for the electrification of the Liverpool, Crosby and Southport line at 625V DC on the 'fourth rail' system, which Dick, Kerr & Co. had estimated would cost £286,724, but it was to be seven months before a contract was signed with that firm. The L&YR Board, at its meeting held on 25th March 1903, gave approval to Aspinall's suggestion that a further number of officers should be sent to America to enquire into and report on the different methods of working as it might affect their respective departments. He had recommended that the party should comprise Josiah Wharton, Goods Manager; George Banks, Assistant Passenger Superintendent & Goods Train Superintendent; F.T. Dale, Assistant Accountant; H.E. O'Brien, Electrical Technician; and James T. Tatlow, Chief Clerk in the Locomotive department. Although Tatlow had previously visited America, Aspinall proposed to send him again to act as secretary to the party.

*White Star liner RMS **Celtic** at Liverpool; this is the vessel on which Eoghan O'Brien sailed to the USA in 1903 with five of his L&YR colleagues. - [From a postcard]*

The party was due to leave Liverpool on 15th April, but their departure was delayed by a day when the White Star liner RMS *Celtic* had a hole knocked in her side on leaving the dock.[47] In a letter to his mother, Eoghan described the vessel as a 'ship de-luxe' and told her that he had the private use of a large state-room, which looked out on the promenade deck, and he added "everything is on a most regal scale". Having called at Cherbourg and Queenstown, the party arrived in New York on 24th April and checked in at the *Waldorf Astoria* hotel. The next day H.E. O'Brien met with Alfred Brotherhood, the Chief Instructor at the Electrical Engineers Institute, and called on Samuel Sheldon at the Brooklyn Polytechnic Institute. On the morning of Monday 27th April the L&YR party called to the offices of General Electric and Westinghouse Electric Manufacturing Company, and in the afternoon they visited the repair shops of the Manhattan Elevated Railway. The next morning, before departing for Baltimore, a visit was paid to the publishers of the *Street Railway Journal*.

The journey to Baltimore included a stop en-route for a few hours at Philadelphia. On Wednesday 29th the L&YR delegation visited the Baltimore & Ohio Belt Line, the first main line railroad electrification in America, which included a 3.7-mile

---

[47] The 21,035-ton RMS *Celtic* was built for the White Star Line by Harland & Wolff at Belfast. She was launched on 4th April 1901, and made her maiden voyage on 26th July 1902. At the time of her introduction she was the first ship built to exceed the gross tonnage of the *Great Eastern*. Equipped as an Armed Merchant Cruiser during World War I she struck a mine off the Isle of Man on 15th February 1917, and on 31st March 1918 was hit by a torpedo fired from U-77 in the Irish Sea. Back in mercantile service, she collided with the SS *Hampshire Coast* in the Irish Sea on 21st April 1925 and with the US *Anaconda* off Long Island on 29th January 1927. She was blown ashore and wrecked at Roche's Point off Cobh on 10th December 1928 and was broken up in-situ by a Dutch salvage firm.

CHAPTER FIVE **Liverpool - Southport Electrification**

tunnel carrying the B&O under part of the city and harbour. Steam locomotives arriving with trains at either end of the tunnel had to bank down their fires while an electric locomotive was coupled on to pull the train through the tunnel. Three 360hp, 600V DC, 96-ton, steeple-cab locomotives had operated successfully on this service since their introduction in 1895. For the next part of their tour the party travelled via Harrisburg to Altoona; a stifling hot journey made none the better by a failed freight train, which caused a long delay. When the party eventually arrived in Altoona at 2 a.m. they found that there was not a room to be had at that hour of the night. However, a friendly railroad conductor found each of them a bed in a deserted hotel adjacent to the freight yard. Although the others could not sleep because of the noise from the whistles and bells of locomotives and of freight cars banging into each other when they were coupled-up, Eoghan slept like a top.

On 30th April the whole day was spent in the Pennsylvania Railroad workshops at Altoona, and those facilities must have seemed enormous when compared with the workshops at Horwich. The party then moved on to Pittsburgh, where they stayed at the *Hotel Duquesne*, and that part of the itinerary offered H.E. O'Brien an opportunity to meet and discuss various aspects of electric railway traction with engineers at the Westinghouse Electric Manufacturing Company; they were supplying alternating current traction equipment to the Baldwin Locomotive Works. On the same day a call was also made to Union Switch & Signal, and a visit to the Westinghouse Air Brake Works at Wilmerding, Pennsylvania, followed during the morning of the next day before the delegation departed for Chicago.

In Chicago the Northwestern Elevated Railroad was visited on 4th May and on the next day the subject of attention was the Metropolitan Elevated Railway. Before the delegation moved on to Cleveland, a day and a half was spent on the Aurora, Elgin & Chicago Railroad. Friday 8th May was devoted to the workshops of the Lake Shore & Michigan Southern Railway at Collinwood, Cleveland, followed by travel to Toledo. On the Saturday the Toledo & Western Railway was visited and a call was made to the Ford Plate Glass Company's power house before the party moved on to Buffalo. A break at Niagara on 10th May ensued, where Eoghan went up to the foot of the falls on board the *Maid of the Mist*, after which he took a trolley car ride down the gorge to Lewiston. There was a visit to the power station, repair workshops, and sub-stations of the International Railway Company on 11th May and to the Buffalo & Lockport Railway and also to the Niagara power house on the following day.

The subsequent overnight journey from Buffalo to Albany presented Eoghan with his first experience of travelling in American sleeping cars, which he found to be most comfortable with a wide, springy bed positioned in such a manner within the car that passengers lay in the direction of travel. On 13th May a call was made to the Albany & Schenectady Railroad and time was also spent at the General Electric

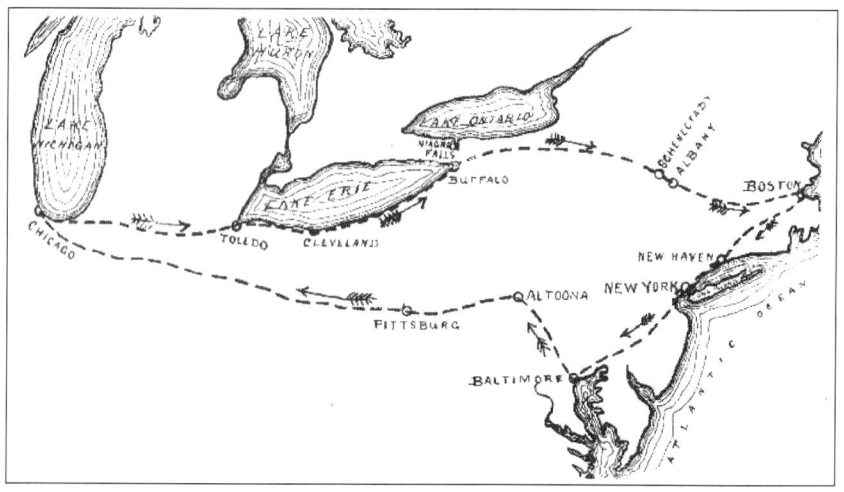

Map showing the route taken by the L&YR delegation on their visit to the USA in 1903 - [From H.E. O'Brien's report]

Company's experimental railway. The following day was spent at the workshops of the American Locomotive Company, Schenectady, where Eoghan was able to obtain information about the direct current electric locomotives that were being constructed there. The production of these locomotives was the result of close collaboration with the General Electric Company which was responsible for supplying the electric traction equipment and specialist technical advice to Alco.

Before travelling to Boston on15th May the delegation called to the Albany & Hudson Electric Railway. The whole of the next day was spent with the Boston Elevated Railway, and a day's relaxation in Boston on the Sunday was followed by a visit to the works of the Thomson Houston Company at Lynn on 18th May. The L&YR party examined the facilities at Boston South Union station on the following morning before departing for Newhaven, Connecticut. The 20th and 21st May were spent with the New York, New Haven & Hartford Railroad, firstly at their Berlin power station, and then at various other facilities before travelling on that line to New York in the evening. The next day included a visit to the offices of the Interborough Rapid Transit Company followed by a farewell call on Mr Brotherhood.

After a free day for relaxation in New York, the itinerary concluded with visits to the power stations, sub-stations and car sheds of the Manhattan Elevated Railway on 25th May and to the Brooklyn Heights Railway on 26th May before the party set sail from New York for Liverpool on Wednesday 27th May aboard the SS *Armenian*, which arrived in Liverpool on 5th June.[48] However, the visit to America was a

---

48 The 8,825-ton SS *Armenian* was a cargo liner built for the Leyland Line in 1895 by Harland & Wolff at Belfast, but the vessel was managed by the White Star Line from 20th March 1903.

couple of years too early to observe the first major commuter railroad electrification in America in operation. This was the 600V DC third rail system implemented by the Long Island Railroad in 1905 that involved procurement of 134 multiple-unit electric passenger cars with steel bodies.

On his return from the visit to America, Eoghan was delighted to learn of his election as an Associate Member of the Institution of Electrical Engineers, Robert Kaye Gray, President of the Institution, announcing this event at the AGM held on 25th May 1903. It was around the same time that he decided to forego membership of the I.Mech.E., preferring instead to concentrate on professional affiliation with the 'Civils' and the 'Electricals'. Just one month later, on 24th June, Aspinall's suggestion of appointing H.E. O'Brien as Resident Electrical Engineer for the Liverpool, Crosby and Southport line was considered by the L&YR Board and, acting on the recommendation of the General Purposes Committee, his appointment was approved at a salary of £200 p.a.. At that same Board meeting the report from the delegation that had visited America was laid on the table.

In January 1903 the L&YR Board had decided that the electrification scheme should be expanded to include the line from Southport to Crossens, which ran through an area with a large residential population. Therefore, the contract entered into with Dick, Kerr & Co. Ltd. on 28th May 1903 was for the complete equipment for electric traction of the Liverpool to Southport and Crossens lines. With certain exceptions, it was intended that the contract should comprise the whole of the expenditure necessary to completely equip the line in accordance with the specifications. It provided for the electrification of about 18½ miles of double track, including the necessary crossover roads at all stations, and an additional two miles of single track in sidings at Liverpool (Exchange), at Hall Road around the fork, and into the carriage sheds at Southport (Chapel Street). The exceptions referred to above consisted of:-

1. The rolling stock, which was to be supplied to the contractors ready to receive the complete electrical power and lighting and heating equipment, and which was to be installed at the company's works as the construction of the coaches progressed.

2. The foundations and buildings for the Power House and the foundations for the substations, including the necessary foundations for all machinery.

3. Any alterations to permanent way found necessary owing to the impossibility of obtaining a continuous contact with the conductor rail in a single length of car.

The contractor was also obliged to start up the line when ready for the public service in line with all conditions that were considered necessary by the BoT.

The contract price of £259,480 was made up as follows:

Power House buildings	£19,429
Power House equipment	£94,626
Rolling Stock equipment	£41,060
Sub-Station equipment	£36,900
High tension distribution system	£27,100
Permanent Way equipment	£40,365

In view of the fact that the L&YR had not fully decided on the extent to which they would fit guards to the live rails, it was agreed that the company would supply all the material for that work and that the contractors would lay the guards for 4d per yard (double guard) and 3d per yard (single guard). This arrangement was based on the understanding that the material would be delivered to the contractors to enable the guards to be fixed at the same time as the third and fourth rails were being laid.

Eoghan was home in Killiney for the weekend of 4th & 5th July 1903 to celebrate his promotion, and on the Sunday afternoon the Smythes were at *Mount Eagle* for tea. On his return to Lancashire he moved into lodgings with a Mrs Wilde at Formby, but unfortunately he was too busy with the new job to join his family in Germany and Switzerland during July when they were accompanied by Olive and Frances Smythe. However, Murrough and Ellen did go over to see him in October, staying at the *Exchange Hotel* in Liverpool, and while there they called on the Aspinalls who had invited them for tea at *Gledhill* on the Sunday.

Electrification work proceeded at a remarkable pace. The three sub-station buildings were completed by 29th July; the Power House building and the high-tension cables by 11th November; and the first high-tension current was produced on 20th December, although complaints had been received regarding the nuisance caused by noise of the steam blowing off at the Power House when the safety valves were tested. The first sub-station was completed, current supplied to the rail, and an experimental run made with one of the new four-car trains on 29th December 1903. Further trial trips were made, including a run from Southport to Formby, on the following day.

It was whilst working on the Lancashire coast electrification that H.E. O'Brien was to become involved with the investigation into the derailment that occurred on 15th July 1903 at Waterloo station, Crosby. The 4.30 p.m. businessmen's express passenger train from Liverpool to Southport, which was booked to run the first 17½ miles from Liverpool in twenty minutes at an average speed of 50 mph, became derailed on the inside of the right-hand curve approaching the station. The locomotive, Horwich built radial tank No.670, which was travelling chimney first, left the rails followed by the whole of the train of six bogie carriages. Having left the rails the locomotive mounted the ramp of the island platform, demolished the supports of the footbridge, wrecked a signal cabin, and jumped completely round to come to a halt facing back to Liverpool with its right hand trailing end hanging over the platform.

Scene of the accident on the L&YR at Waterloo station on 15th July 1903 - [Author's collection]

The locomotive and the two leading carriages collided, which wrecked the passenger compartments of the carriages and resulted in fatal injuries. Five passengers and the fireman died at the scene and one-hundred and twelve passengers, the driver and three other servants of the Railway Company were injured. Those who lost their lives were: Edward Rigby, the fireman; Miss Donna Maria Walters; Mrs Fanny Ryder and her daughter Florence; Miss Edith Morgan; and Mary Latham. One other passenger died later from injuries received in the accident. No.670 was quite badly damaged and bogie Third Brake No.2963 had its body, underframe and bogies smashed. Bogie Third No.1207 also had its body, underframe and bogies smashed and bogie First Class No.43 had two end panels broken and other panels badly damaged. Bogie Composites Nos.421 and 665 also received body damage and the last vehicle, bogie Third Brake No.2908, had just one bogie off the road and was more or less intact.

Nothing was found indicating that any obstruction existed on the line near the point of derailment, and the permanent way was, as far as could be judged, in good order. On examination of the line after the accident a bridle from the spring of the right trailing coupled wheel of the locomotive was found in the 6-ft space between the up and down lines, close to the inside rail of the down line at a distance of 65 yards from the footbridge. Some 19 yards nearer the station the whole spring from the right driving wheel was found, also close to the inside rail, nearly buried in the ballast. The pin was broken and had apparently been run over, and the spring bridle was broken off close to the bottom of the yoke, with adjusting pin still in it. The spring was practically undamaged, and neither bridle nor spring had been run over. Springs on the axles of the coupled wheels of the radial tanks were low down under the axle boxes, and they were carried in yokes that were attached to the axle boxes by a yoke pin. The springs were in two portions, the top half pressing against part of the engine frame and the lower half against the bottom of the yoke called the bridle, in which there was a screw pin for adjustment. The springs were carried so low down that, if a locomotive was derailed, the head of the adjusting pins would touch the rails when the flanges of the wheels were on the chairs.

The Inspecting Officer who conducted the formal inquiry for the BoT was Major Edward Druitt, RE.[49] In his evidence to the inquiry, Henry Hoy stated that he had never known one of these springs to break, but if such were to happen he stated that the engine would yaw about to a dangerous degree. It was his opinion that the engine had run over some obstruction, either something lying on the rails or some-thing that belonged to itself. He did not think it possible for the broken pin to have derailed the engine had it run over it. Driver William Lloyd stated that he had the engine in mid-gear and just approaching the footbridge at Waterloo he felt something wrong with the engine. He stated, "It gave a sudden leap up as we were approaching the footbridge and left the rails. There was no special yawing motion, the engine simply seemed to jump and mount up bodily — it lifted all of a sudden."

Major Druitt considered that the dropping of the spring of the right trailing coupled wheel must be regarded as the most probable cause of the derailment, as judging from the damage to the permanent way he did not consider that excessive speed alone could account for the derailment, for had the engine mounted or burst the outer rail, it and the carriages behind, would have been derailed on the outside of the curve, whereas the contrary was the case. The curve in question was graduated, starting at 40 chains radius for a length of 80 yards, before changing to 23 chains radius, with a maximum super-elevation of 2½in. The Inspecting Officer considered that a speed of 50 miles an hour round such a curve was not desirable, and recommended that, as then constructed, speed on it should not exceed 35 miles an hour.

Samuel Brighouse, the Coroner for West Lancashire, held an Inquest into the accident at Waterloo town hall on 4th August. The jury found that the deceased persons were accidentally killed by the derailment of the train, but that the evidence produced did not enable them to arrive at the precise cause of the accident. Although the cause was not fully explained at that time it was generally attributed to a spring defect causing the weights on the coupled wheels to go awry; it was not attributed

---
49 National Archives, PRO, MT1053/92 BoT Returns and Reports of Railway Accidents, 1903

CHAPTER FIVE **Liverpool - Southport Electrification**

to a design defect. However, H.E. O'Brien was later to recall, "It took us a long time to discover the cause of the derailment".[50] Hoy, recalling O'Brien's prize-winning paper on bearing springs, put him to work on the problem with J.R. Billington, a draughtsman that Eoghan described as 'really brilliant'. The problem was tracked down to a small design fault, which was quickly corrected. As a result of Billington and O'Brien's findings, under-hung plate springs were adopted from June 1904 onwards in place of the double elliptical springs that Aspinall had previously used on his locomotives.

Henry Hoy had resigned his position as CME of the L&YR on 8th March 1904 in order to accept the role of Managing Director with the locomotive builders Beyer Peacock & Co. Ltd., Gorton. A series of consequential promotions took effect from 12th March which saw George Hughes become CME and Oliver Winder Assistant CME & Works Manager, Horwich. At the same time Winder was given a special grant of £150 for his services supervising the overall electrification project, including the provision of electric rolling stock. Nigel Gresley took over from Winder as the Assistant Carriage & Wagon Superintendent; C.H. Montgomery filling the post of Works Manager, Newton Heath, vacated by Gresley, and Frederick William Attock replacing Montgomery as Outdoor Assistant in the C&W Department.

Throughout the first months of 1904 Eoghan was very busy, especially on Saturdays and Sundays when trial trips were run. It was during this period that Richard Cronin, Locomotive Superintendent of the DW&WR, visited the railways in England that were adopting electric traction. Amongst those he examined in London were the Central London Railway, the Metropolitan District Railway, the Metropolitan Railway and the Waterloo & City Railway. In Liverpool he visited the Liverpool Overhead Railway and also the Liverpool and Southport line of the L&YR, where electrification of the latter was nearing completion, and his former pupil, Eoghan O'Brien, afforded him every facility. The L&YR system appears to have impressed Cronin the most and in his report he enthused "Perhaps this is the most important and interesting line I visited... ...and it approximates more nearly to our own line and conditions than any other line I saw".[51]

On 4th March 1904 a trial run was made over the whole 21 miles from Crossens to Liverpool which, at that time, was the longest non-stop run that had ever been made by an electric train in the U.K. An express run from Liverpool to Southport, which reached a speed of 70 mph, was made on 20th March and two days later the official 'first electric train' left Liverpool Exchange station at 11.35 a.m. with Aspinall at the controls. The Crossens line was opened to electric traction on 22nd March 1904, and a mixed electric and steam service was operated from 5th April on the Liverpool Crosby and Southport section until 11th April when a subsidence

---

50 *Horwich & Westhoughton Journal*, 29th September 1967
51 *Report on Visit to Electric Railways in London and Liverpool*, Richard Cronin, DW&WR, 1904.

at Formby Power House — which, due to the marshy site conditions, had been constructed on a substantial concrete raft — forced a return to full steam working. The problem was quickly rectified by Monk & Newells, the contractors for the Power House foundations, and mixed electric and steam services were resumed on 13th May. However, it was not until October 1904 that sufficient electric rolling stock had been delivered to enable the full electric service to be provided.

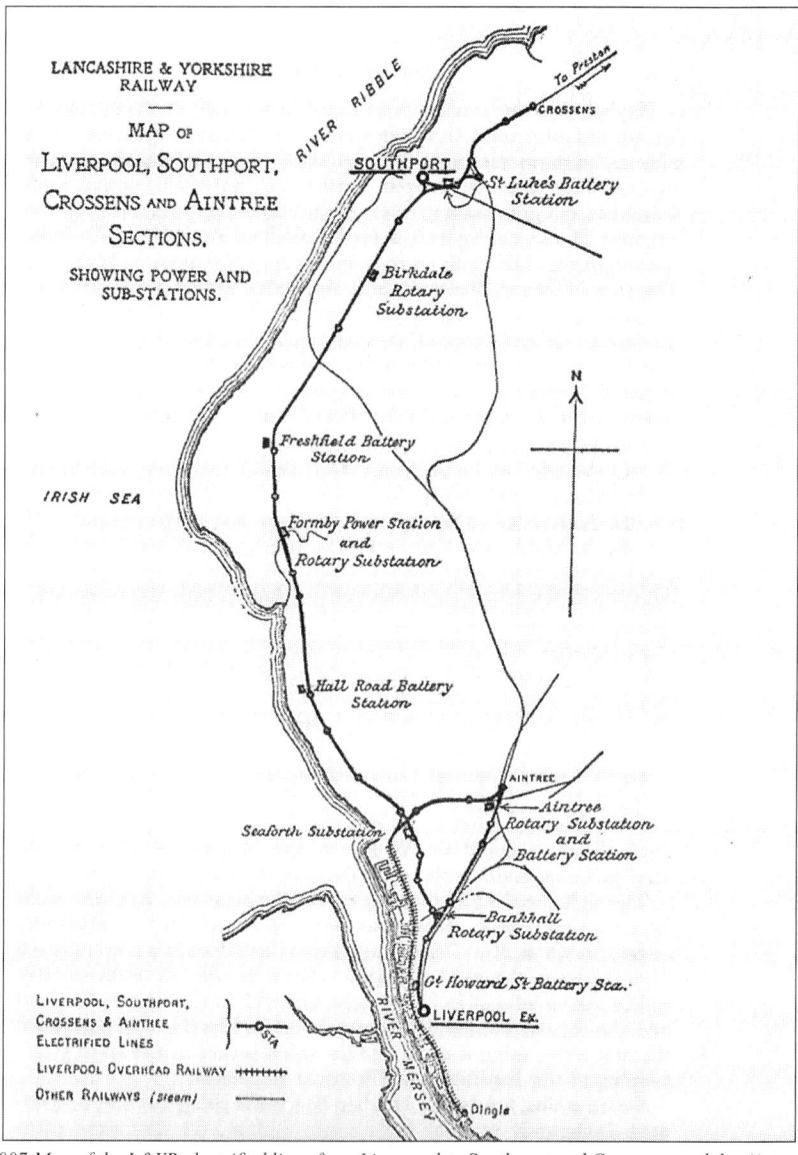

1907 Map of the L&YR electrified lines from Liverpool to Southport and Crossens, and the Aintree extension, showing the Power Station at Formby, four rotary sub-stations, and five battery stations.

Murrough went over to visit Eoghan for the weekend of 23rd-24th April and found him still very busy. He only had time to look in on his father for five minutes each morning and would spend all day at the Power House, sometimes not even getting home in time for dinner, although he did manage lunch with his father on one day. They dined together one evening and had soup and sole "with a bottle of the best wine his father could select," which Eoghan seemed to enjoy. With the commencement of the electric service, the Board approved H.E. O'Brien's appointment as Engineer for Electric Traction on 26th April 1904 and at the same time his salary was increased from £200 to £300 per annum. In advising him of the news, George Hughes added in his covering letter, "I hope the enclosed will give you as much pleasure as it gives me in forwarding".[52]

Based at Formby, Eoghan had full charge of the Electric Traction Department. In addition to the operation and upkeep of the Power House, he was responsible to the CME for the operation and maintenance of sub-stations, track equipment and rolling stock. Some idea of his wide range of responsibilities can be gathered from the following descriptions of the generating plant and equipment and the electric rolling stock employed. The Power House, which provided the electric power for traction purposes, was located adjacent to the railway, on the banks of the River Alt, at Formby, midway between Liverpool and Southport. Water from the River Alt was used for condensing purposes, avoiding the need to construct cooling towers. Each main engine unit was provided with an independent jet condenser with air pumps of the Edwards type driven from the tail rods of the horizontal engines. For the exciter units and subsidiary plant a surface condenser with combined air and circulating pumps was installed. Boiler feed water was drawn from a well provided by the contractors. Three feed pumps of the Mather & Platt type were provided, any one of which had ample capacity to deal with the normal load of the station, and the feed water piping was arranged in duplicate throughout.

The Power House building was 290ft long by 130ft wide designed as an independent steel structure, the stanchions of which, in addition to carrying the roof, also provided support for a 20-ton electric travelling crane over the engine room. Spaces between the stanchions were filled with brickwork. There were two spans, one containing the boiler plant and the other, larger, span housing the engine-generator plant, including steam-driven exciters, sub-station plant and the main switchboard. Two ornamental iron stairways provided access to the switchboard gallery, which was surrounded by an ornamental hand railing. The switchboard contained all the equipment for handling the distribution of power, not only to the substations but also for the lighting and other supplementary plant throughout the Power House, together with all of the necessary appliances for ascertaining the output and efficiency of the generating plant.

The boiler plant comprised sixteen Lancashire boilers, each 32ft x 8ft 6in. diameter, designed for 160 lb/in^2 working pressure, with furnaces 3ft 5in. diameter arranged

---
52 Letter dated 30th April 1904 from George Hughes to H.E. O'Brien

for hand firing. The bunkers held about 145 tons of coal, and mechanical stokers were subsequently provided. Each boiler was equipped with an independent superheater of the Galloway type placed in the main flue. The boilers were arranged in separate batteries of four, each of which, under ordinary conditions of working, was connected direct to one of the main units, but each battery could also be connected to a common steam main, thus allowing for any battery being connected to any engine. The main steam pipe ran the whole length of the boiler house and was connected to each engine and to each battery of boilers, thus forming a duplicate supply system.

Lancashire boilers (l) and cross-compound horizontal reciprocating engines (r) at Formby Power House - [HOR-F-607 and HOR-F-604 © NRM, York]

The products of combustion from the boilers were dealt with by two induced draught fans, each fan being capable of dealing with the gases from 16,000 lb of coal per hour, or the equivalent of 12 boilers, so that under normal conditions only one fan needed to be in operation. These fans were driven by independent direct-coupled engines running at about 175rev/min. Each unit was complete with the necessary flue connections, dampers and 6ft. 6in. diameter, 60ft high wrought iron chimneys. The ash handling plant was a narrow gauge tramway system with turntables, etc., located underneath the firing floor. Twelve narrow gauge steel tipping wagons were used to transport the ash and a motor driven hoist was employed to elevate these wagons to a suitable height for tipping into railway trucks.

The generator plant consisted of five main units. Four of these units were driven by horizontal cross-compound condensing type marine engines having cylinders 32in. and 64in. diameter by 54in. stroke and running at a speed of 75 rev/min. Both cylinders were fitted with Corliss valves, the exhaust valve in each case being operated by an independent eccentric. The large flywheels were coupled direct to the spiders of the generator rotors. The 3-phase generators rated at 1,500kW produced high tension current at 7,500V with frequency of 25Hz. The fifth unit, a 750kW alternating current generator, was driven by a vertical compound Corliss engine, having cylinders 23in. and 46in. diameter by 42in. stroke, and running at a speed of 94rev/min. Three independent steam-driven exciters were installed, each with ample capacity to provide excitation for four 1,500kW capacity units and sufficient margin for all lighting and motors in the Power House.

In addition to that located in the Power House, sub-stations were provided at Bank Hall, Seaforth, and near Birkdale. The equipment for the sub-stations at Bank Hall, Seaforth and at the Power House comprised four 600kW rotary converters, three of which were capable of dealing with the service of the line during the heaviest hour, leaving one as a spare unit. At Birkdale sub-station there were three 600kW rotary converters, two of which were capable of dealing with the service of the line during the heaviest hour, again leaving one as a spare unit. At each sub-station three static air-blast transformers were provided for each of the above units and an additional one was installed to take the place of any one that might break down. Two motor driven fans were provided for supplying the air blast to the transformers, each of which was capable of dealing with the full-load requirements of four rotary converters. The 7,500V 3-phase, AC supply to the sub-stations was converted and stepped-down to 650V DC for traction purposes by the rotary converters and static transformers.

Generators and Control Gallery at Formby Power House - [HOR-F-4331 © NRM, York]

The switchboards at the sub-stations were arranged to receive the 7,500V high tension alternating current and distribute it to the rotary converters and control the 650V low tension direct current and its distribution. In starting the station under normal conditions, current was taken from the third rail which had already been charged from the Power House. The distribution of high tension current from the Power House to the sub-stations was by means of three-core paper insulated metallic-sheathed cables suitable for 10,000V. The cables were arranged so as to allow for the service being continued should a fault occur in any one cable, thus constituting a triplicate system of supply. The cables were laid on what was known as the 'solid system', generally in trenches in the 6-ft way that gave a clear 18in. above the top of cables. Troughs made of creosoted wood were used to protect the cables with the entire covering filled in with insulating compound.

The live conductor rails were supported on insulators located on the outside of the running rail at a distance of 3ft 11½in. from the centre line of the track, and the top surface of the conductor rail was 3in. above the top of the running rails, being the dimensions agreed upon between the British railway companies adopting third rail electrification. The supporting insulators were located generally at 10ft intervals on every fourth sleeper. Conductor rails were of Vignoles section and of an average resistance equal to not more than 7¼ times that of pure copper. Ordinary joints were close butted and provided with two-bolt fishplates. An expansion joint was provided approximately every 300ft. The ordinary joints were bonded with laminated copper bonds having a combined sectional area of $0.8in^2$. Bonds at the expansion joints were of the same area, but of a suitably flexible design. Where it was necessary to break the run of the conductor rail, either at level crossings or at track junctions, the rail ends were ramped down and the electrical continuity was completed by means of insulated cable laid either in the clearway or underneath the sleepers.

A return conductor rail, which was of a similar composition to the live conductor rail and bonded in a similar fashion, was laid between the running rails and bonded to each of them. The return current from the motor cars was taken through the running rails into the return conductor rail. The connection between the two conductor rails and the sub-stations was by insulated cables connected to the rails by special copper contact plates. The 650V DC 'fourth rail' system adopted by the L&YR avoided the need to bond the running rails and provided a more certain return path; it also became the standard arrangement for the London Underground electrified lines. Large battery stations were constructed at Great Howard Street (near Liverpool Exchange), Hall Road, Freshfield and St. Lukes (Southport) in 1905 to secure: (1) a constant load at the main generating station when severe peak loads were placed on the system; (2) a steadier traction voltage on the insulated rail; (3) an increase in the spare capacity available at the generating station; and (4) a valuable standby in the event of a partial and temporary failure in the supply of current.

The original service was designed around twelve four-car train sets in operation at peak hours and the contract with Dick, Kerr & Co. included a sum of £41,060 for the supply of all electrical equipment for twenty-eight motor cars and twenty-six trailer cars. Each set comprised a third class motor car at each end between which were sandwiched two first class trailer cars. The total number of cars allowed for one set being out of service for maintenance purposes and there were two additional spare motor cars. The cars were constructed at Newton Heath during 1904, and Beyer Peacock manufactured the 8ft wheelbase motor bogies, to a design developed by Henry Hoy incorporating equalising beams.

The new electric vehicles were built to a distinctive design, their general appearance and interior layout following American rather than British practice. Each car was 60ft. 4¾in. long with a body width of 10ft 0in. — the widest ever used in Britain. Body construction followed the American pattern and was extremely strong; exterior panelling being finished in the matchboard style. They had clerestory roofs, and the passenger compartments were of the open saloon style with centre aisles. The third class motor cars (built to L&YR Diagram 56) were equipped with four Dick, Kerr 150hp traction motors and the power control system was by direct control of the motors, an arrangement which did not permit two sets to be worked together in multiple. The trains were equipped with vacuum brakes, which was unusual for electric stock. Whereas the outer ends of the motor cars were provided with normal buffers and couplings, the intermediate connections between the cars in each set was by means of automatic couplers. The passenger entrance doors were recessed so that, although opening outwards, they did not project beyond the sides of the vehicles — a feature that provided a level of safety to passengers standing on platforms that was not available with the ordinary passenger stock.

Official L&YR photograph taken at Newton Heath showing first of the 1904 L&YR 4-car electric sets for the Liverpool - Southport service - [HOR-F-123 © NRM, York]

The third class motor cars provided seating for 60 passengers (later increased to 80) and the first class trailer cars (built to Diagram 57) had 66 seats, thus each set originally provided seating accommodation for 132 first class and 120 third class passengers. The electrification proved so successful, and the consequent growth in passenger numbers was so rapid, that it soon became necessary to increase the formations to five-car units. An additional two third class motor cars and two first class trailers to the same design were constructed in 1905. At the same time it was also recognised that the proportion of first class seating was too high in the original sets, so it was decided that the twelve additional trailer cars to be built during 1905 would be third class vehicles (Diagrams 68 & 69), which provided seating for 80 passengers. Further details of the electric stock for the Lancashire coast electrified lines are presented in Appendix Two.

Front end of Electric Motor Car showing taper for clearance (l) and electric coach end showing inset passenger entrances, power and lighting control jumper cables, and automatic centre coupler (r). - [HOR-F-126 and HOR-F-127 © NRM, York]

Maintenance and repair of the electric trains was undertaken in the former carriage maintenance shed at Southport, Chapel Street, which had been specially converted for the purpose. This workshop, which was situated on the up side of the railway in close proximity to the station, was 272ft long by 55ft wide and had three roads with standing room for twelve 60ft cars. Two roads were equipped with inspection pits and a wagon turntable on each road enabled bogies to be transferred from one road to another. Overhead cranes were provided for the moving about of armatures and light material and there were fixed jacks in each pit for dropping out the traction motors. The back end of the main shop was reserved for dismantling and assembling bogies. A small machine shop adjoined the main building and was equipped with a drilling machine, a planing machine, and two lathes which were used for skimming traction motor armatures and turning up journals, etc. The work control system included the use of component record cards, indicating the length in service and number of miles run, which were attached to each bogie and each traction motor undergoing repair.

CHAPTER FIVE **Liverpool - Southport Electrification**

Motorman's cab in 1904 Motor Car (l) and original power bogie for 1904 electric stock to a design developed by H.A.Hoy (r). - [HOR-F-128 and HOR-F-129 © NRM, York]

Interior of L&YR 1904 Liverpool - Southport electric stock: First class (l) and Third class (r) - [HOR-F-124 and HOR-F-125 © NRM, York]

On 2nd May 1904 Eoghan sent a telegram to his parents telling them of his £100 rise in salary and that he had been granted three days leave, and letting them know that he would arrive at *Mount Eagle* on the following morning. On his first day at home he rode with his father and the Smythes to Kilmacanogue and they all walked up the 1,654ft Great Sugar Loaf. The next day they went to the waterfall at Powerscourt, and Mrs Smythe, together with Frances and Barbara, rode with Eoghan. The three days passed quickly, but he was in possession of the appropriate means of supporting a wife and took the opportunity to discuss future plans with Frances. He was back again at Killiney on 27th July for two weeks holiday. On the evening of 4th August the family were in the drawing room at *Mount Eagle*; Murrough talking business with a certain George Simms and aunt Kate Toulmin and her daughter Norah working on lace with Ellen, when Kate's maid came in and said, "Mr Hohen wishes to speak to you". Ellen and Murrough went out and there was their son with Frances, who was blushing and smiling and half-crying with excitement and joy, as Eoghan announced the news of their engagement.

79

Life with an Electric Railway

The foremost thing on Eoghan O'Brien's mind when he returned to Formby in August 1904 was the question of finding a house to live in and for that purpose he secured a lease on *Granagh*, a dormer bungalow situated in Andrew's Lane. Ellen went over a couple of weeks later with "all kinds of resources of civilisation" to help him get the bungalow prepared. Eoghan met her at Liverpool and took great delight in bringing his mother out to Formby on one of the new electric trains. Murrough brought Frances over on 27th September for her first look at *Granagh* and she stayed there with Ellen until 18th October. Everyday they bought some little addition for the house, and when they had all departed, Eoghan settled in at the bungalow. He was not able to go home to Killiney for Christmas 1904, as he was busy working on the new electrified section that would connect the Southport line with Seaforth Sands on the Liverpool Overhead Railway, and also because he required time off for his wedding in January.

The wedding of Henry Eoghan O'Brien and Frances Victoria Lucy Smythe took place on Tuesday 10th January 1905 at Holy Trinity Church, Killiney; the service being conducted by the Rector, Canon R.B. Stoney, and Revd Dudley Fletcher, incumbent of Lissadell, Co. Sligo, and brother-in-law of the bride. The best man was Ventry Guiscard Mellin, a friend of Eoghan's in the traffic department of the L&YR. The groom's party travelled in a Brougham drawn by a splendid pair of horses up the hill from *Mount Eagle* to the church, which was beautifully decorated with palms and arum lilies. The bride, who wore a white crepe de chine dress with chiffon flounces embroidered in silver sequins and a veil of rare old Limerick lace lent by her mother, was given away by her father, R.A. Smythe. The bridesmaids were Olive Mary Smythe and Eileen Barbara Smythe (sisters of the bride), Miss Irene Falkiner, and Erica Smythe a six-year old cousin of the bride. They carried bouquets of lilies, and wore pearl brooches that were gifts from Eoghan.

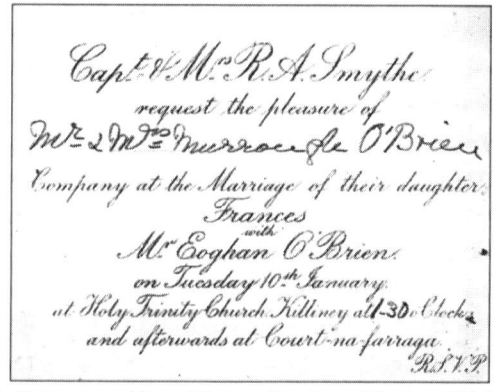

Invitation to wedding of Frances Smythe and Eoghan O'Brien
- [Courtesy Prof. O.F. Robinson]

CHAPTER SIX **Life with an Electric Railway**

Wedding at Killiney on 10th January 1905, l-r back: Barbara (Bay) Smythe, Eoghan O'Brien, Frances O'Brien (née Smythe), Murrough O'Brien, Olive Mary Smythe, and Ventry Guiscard Mellin; in front: Irene Falkiner and Erica Smythe. - [Courtesy Prof. O.F. Robinson]

Richard Altamont Smythe (l) and Frances Anne Jane Smythe (née Bellingham) (r), Eoghan O'Brien's parents-in-law - [Author's collection]

A large-scale reception was held at *Court-na-Farraga* where the guests included Sir Henry & Lady Bellingham, Gen. & Hon. Mrs Waller, Rt.Hon. Frederick & Miss Wrench, Sir John & Lady Franks, Mr & Mrs Clifford Lloyd, Sir Frederick & Lady Cullinan, Lady Anne French, and the Countess of Portarlington. The couple received about 180 gifts including a silver mounted umbrella and a copy of Joyce's

Social History of Ireland from Mr & Mrs Michael Davitt and a silver coffee pot and stand from Mr & Mrs John Dillon. Amongst the gifts from railway colleagues was a silver model of an electric motor car controller from J.A.F. Aspinall for the groom, and a bracelet from Mrs Aspinall for the bride. Mr & Mrs Hoy gave the couple a silver egg stand, Mr & Mrs Winder a watercolour, and Mr & Mrs Gresley gave them silver bon-bon dishes. The staff of the Liverpool & Southport electric branch presented a silver tea service.

The silver model of the driver's controller used in the 1904 L&Y electric motor cars, which J.A.F. Aspinall presented to H.E. O'Brien on the occasion of his marriage. - [Prof. O.F. Robinson]

Upon leaving for honeymoon in the south of England, the bride wore a Royal blue costume faced with white and silver. Having crossed to Holyhead on the 7.0 p.m. Mail steamer, Eoghan and Frances spent the next night at the *Charing Cross* hotel before proceeding to Torquay where they stayed at the *Imperial* hotel. They spent their days sailing in the morning and walking in the afternoon and, after a week at Torquay, they went on to Boscombe and finally spent a weekend at Bournemouth before their return to Formby. The couple settled in at *Granagh* and were soon 'snug as two trout' in their little house. Eoghan was kept busy with late night call-outs to the Power House and he was often there until 1.0 a.m. In a three-week period there were four or five serious incidents with the cross-compound engines that resulted in smashed parts, none of which were the fault of the L&YR or its staff.

Then came a surprise announcement; H.N. Gresley resigned his post with the L&YR in mid-February in order to take up the position of Carriage & Wagon Superintendent with the Great Northern Railway at Doncaster. The resulting vacancy was filled by promoting J.P. Crouch to Assistant Carriage & Wagon

CHAPTER SIX **Life with an Electric Railway**

Superintendent; F.W. Attock taking over as Outdoor Assistant in the Locomotive Department and his place being taken by Frederick Stanton Barnes as Outdoor Assistant in the Carriage & Wagon department. It may seem that H.E. O'Brien had been overlooked at that time, but his responsibilities in connection with the L&YR electrified services were increasing with its incremental growth. On 25th July 1905 the connection from Seaforth & Litherland to Seaforth Sands on the Liverpool Overhead Railway (LOR) was opened and a through service of trains from Dingle to Southport via the LOR and the new connection began on 2nd February 1906. However, the latter was short-lived, services being withdrawn on the outbreak of war in August 1914.

***Granagh**, the dormer bungalow, at Andrew's lane, Formby, which was Eoghan and Frances' first home. Frances (l) and Bay Smythe (r) are at the front door. - [H.E. O'Brien; courtesy Prof. O.F. Robinson]*

*Frances O'Brien and Bay Smythe at **Granagh** in 1905. - [H.E. O'Brien; courtesy Prof. O.F. Robinson]*

On the family front Frances' first cousin, Augusta Mary Monica Bellingham, had become the Marchioness of Bute on 6th July 1905 upon her marriage to Sir John Crichton-Stuart at Castlebellingham, Co. Louth. On 23rd of the same month, at the beginning of a two-week visit to *Mount Eagle* with Eoghan, Frances developed Scarlet Fever and had to be isolated, and a resident nurse was called in to attend her. Four days later Eoghan received a telegram requesting him to return at once, as there had been another serious railway accident on the Liverpool – Southport section near Crosby, which by its gravity eclipsed the events at Waterloo just two years earlier. This was the collision that occurred on 27th July 1905 at Hall Road station when the 6.30 p.m. Liverpool to Southport express passenger train collided with an empty passenger train standing in the middle siding. Both trains were five-car electric units, each with seating for 350 passengers.

The collision was a very violent one, the leading car of the express and the first car of the empty train were telescoped together; both bogies of the first car of the empty train being driven back underneath the second car of that train, so that there were four bogies underneath it. The force of the collision drove the empty train back a distance of about 50 yards, but the only wheels that became derailed were

83

the leading pair on the front bogie of the express. There were 56 third-class and 20 first-class passengers in the express and, of those passengers in the leading car, 20 died at the scene and two were severely injured, one of whom died later. The motorman of the express, William Rimmer, was also severely injured, including a blow to the head that rendered him unconscious for five days after the accident, and 45 other passengers complained of minor injury or shock. No.3023 and No.3015, the respective third class motor cars of the express and empty train, had their bodies demolished and their underframes badly damaged at one end. The motor bogies were badly damaged, with axles and frames bent, but only one motor suffered any damage. The electrical and other equipment in the motormen's compartments was destroyed and other electrical equipment under the cars was extensively damaged. The front ends of the second coach in each train — No.401 in the express and No.3105 in the empty train — were smashed in.

Hall Road accident 27th July 1905; Third Class Trailer Car No.3105 resting on two power bogies and two trailer bogies as a result of the force of the collision. - [Author's collection]

At Hall Road Station there were two facing connections in the down line to the north of the station, one to the left that led to the down sidings, and one to the right that led to the middle siding which was situated between the Up and Down lines. The inner home signal, located two yards before the facing points to the down siding, had three arms. The centre one for the mainline was a full-sized arm and higher than the two smaller arms for the sidings. In addition to the mechanical locking of the levers in the frame of the signal box, the facing points were equipped with mechanical detectors making it impossible to pull off the inner home signal for the mainline unless both sets of facing points were correctly set for that line.

The local train, which departed from Liverpool Exchange station at 6.20 p.m., arrived at Hall Road station at 6.38 p.m. and, after discharging its last passengers, it moved forward through the facing points into the middle siding where it came to a stand at 6.40 p.m. Once the empty train was in the middle siding, W. Boote, the signalman who was on duty in the Hall Road signal box, reversed the facing points, but when he tried to lower the inner home signal for the express he was unable to do so. Believing that the facing points to the middle siding were not truly home he worked them again three times in an effort to move them into their true position. Unfortunately, Boote finished up with the points set for the siding instead of the mainline.

Finding that he still could not lower the inner home signal he went to the window of his signal box and showed a green flag to the motorman of the express to signal the train to pass the inner home signal at danger. Motorman Rimmer was running slowly, but on seeing the green flag, he put on power and was running at 40 to 50 mph when he saw that the facing points were set for the middle siding and not for the mainline. Judging by the horrific result of the collision, the speed of the express train could not have been reduced by much before the impact and after the accident the handle of the power controller was actually discovered in the full-on position. Rimmer probably made a full brake application, as evidenced by the circuit breakers in the motorman's compartment having come out due to an exceptional amount of current draw.

Once again, it was Edward Druitt, who conducted the formal inquiry for the BoT.[53] He considered that the collision was due solely to signalman Boote having inadvertently left the facing points set for the middle siding, which he compounded by neglecting the rules that required him to send a man to see that the points were in the correct position. Lieut-Col. Druitt also pointed out that motorman Rimmer should have stopped at the signal box to ascertain what the signalman meant by showing him the green flag. Looking at the facts, he felt that the two most striking features were, first, the absence of a fire, the danger of which was considered as one especially likely to occur on electrified lines; and, secondly, the very small amount of damage done to the passenger cars (except for the leading ones), which he felt was due to the heavy and stiff underframes employed in their construction. The Hall Road collision attracted a lot of attention owing to the high number of deaths and to the fact that it occurred on an electrified section of railway.

The Hall Road collision resulted in Eoghan having fewer vehicles to operate with, but he still had to meet the daily requirements of the full service schedule. The two damaged trailer cars were repaired at Newton Heath, but the two motor cars were so badly damaged that they were considered to be beyond economic repair, and it was to be nearly a year before two new replacement vehicles were completed. The Chapel Street maintenance depot was obviously under pressure and, following a conversation that had taken place on the previous Saturday, Oliver Winder wrote on 14th August confirming his anxiety in regard to cutting down on expenditure for the half-year, both in labour and materials. Winder pointed out several ways in which it would be possible to economise on material and stressed that a strict watch should be kept on overtime. He asked O'Brien to report within a fortnight on what he had been able to achieve and to make recommendations compatible with safety and efficiency.

When Frances returned to Formby on 7th September, Eoghan took a week's leave and the couple went to the Lake District for a break. They stayed at the *Old England* hotel at Windemere, Murrough and Ellen joining them there on 12th September in time for a family celebration of their 32nd wedding anniversary on the following

---
53 National Archives, PRO, MT1053/94 BoT Returns and Reports of Railway Accidents, 1905

Saturday. It was at about that time that the L&YR Engineering & Scientific Club was formed, the inaugural meeting of which was held on 11th October 1905, and Eoghan O'Brien was elected its Vice President. He served the Club in that role from its inauguration until the end of the 1909/10 session. The year 1905 ended with a Christmas surprise in the form of a letter from George Hughes who wrote to him on the 23rd December advising of the Board's decision, taken three days earlier, to increase his salary to £350.

The introduction of the through Dingle to Southport service on 2nd February 1906 has already been mentioned. This hourly service was operated by a fleet of twelve 45ft lightweight electric motor cars working as single cars or in two-car sets. A prototype vehicle, No.1000, had been delivered by Dick, Kerr & Co. in 1905, and eleven other cars, Nos.1001 to 1011, were completed at Newton Heath in 1906. Increased traffic led to the equipping in 1906 of the other two tracks on the four-track section between Sandhills and Seaforth & Litherland for electric traction. During March, Eoghan spent a week in Germany with Oliver Winder, principally to deal with question of a turbine driven alternator set for the Power House at Formby, and they visited Siemens in Berlin and companies in Essen, and Düsseldorf.

The section of the North Mersey branch between Marsh Lane Junction and Aintree was also electrified in 1906, and new stations were built at Linacre Road and Ford. At the same time the short link between the North Mersey branch and the LOR at Rimrose Road Junction, was also electrified. The L&YR commenced operating electric services from Aintree to Dingle on 1st June, but these ceased in September 1908. During this period Eoghan used to go to the Aintree race meetings, or at least to the Grand National, often travelling in the cab of the motor car. However, in later years LOR special trains ran via Rimrose Road Junction over the North Mersey line to Aintree in connection with the Grand National. The direct route from Sandhills to Aintree was also electrified in 1906, electric services being introduced on that line from 19th November.

Additional electric stock was constructed at Newton Heath for the services over the two routes to Aintree comprising of eight third class motor cars (Diagram 73), six first class trailer cars (Diagram 74) and six third class trailer cars (Diagram 81). The first two third class motor cars were actually completed in late 1905, but the remainder of this stock was delivered during 1906. This fleet provided for six sets, comprising a third class motor car at each end between which were sandwiched two trailer cars, one first class and one third class; and two spare third class motor cars. Two additional third class motor cars to Diagram 73 were built in 1906 to replace Nos.3015 & 3023, which had been wrecked in the Hall Road collision the previous year.

During 1906 Eoghan also contributed to the discussions on two papers read before engineering institutions. Following T.H. Schoepf's reading of his paper *Single Phase Railway Motors*, before the Manchester local section of the I.E.E. on 27th February, a spirited discussion took place. A Mr Cramp suggested that the high tension mains could be connected to the field excitation coil of the motor, but H.E. O'Brien felt that the man who would be responsible for repair and maintenance of such motors, with tappings on the 6,000-V stator winding, "would be likely to go stark, staring mad". He went on to question whether the author used air-cooled transformers, to which Schoepf in his reply stated that air-cooled transformers were cheaper and equally as good as oil transformers on voltages up to 6,000, but above that he thought that oil transformers should be used as they were less liable to absorb moisture because of their superior insulation.

Commenting on Charles Frewin Jenkin's paper *Single Phase Electric Traction*, read before the I.C.E. on 13th November, H.E. O'Brien thought that discussions on electric traction as applied to railways rather resembled the consultations of doctors round the bedside of a patient: "the doctors were more occupied in discussing the relative merits of their own remedies than in diagnosing the patient's case." The patient, in question was the railway, and he stated that he desired to say a few words from the railway point of view. Looking at the problem of electrification from the standpoint of efficient railway working he felt that there was little to urge against the single-phase system as its efficiency was practically equal to that of a continuous current system, allowing for the fact that rotary converters were dispensed with, and that the single-phase motors were slightly lower in efficiency than the continuous current motors. But, in installing overhead equipment on mainlines, particularly in the neighbourhood of large towns, he considered that very serious difficulties would be met with, which would entail a large amount of costly special work, causing the total capital cost of the single-phase system to exceed the cost of the continuous current system; expenditure on the alteration of existing structures alone would be considerable.

H.E. O'Brien did not feel that there would be any great difference in the working costs of the two systems, but considered that maintenance of high-voltage overhead equipment would be difficult. He asked those who had studied the economics whether single-phase traction would be applicable to goods traffic on mainline railways, and what arguments could be brought forward to show that such working would be economical. O'Brien felt that single-phase electrification would be a likely method for cheaply equipping extensions to existing continuous current lines, provided that there were few junctions and overhead bridges such that little special work would be required. He also thought that in almost all cases the comparative cost of installation and comparative ease of maintenance of the track equipment would be the deciding factors. He stated that to equip one mile of track with junctions using a 70lb third rail, including all bonding, would cost about £800, but he had not seen estimates for equivalent overhead high-tension construction, which he considered would not cost less than £1,400 per mile.

In reply, C.F. Jenkin stated that it would be difficult to imagine a cheaper construction than wires stretched between light lattice girders and he felt that the work at bridges was almost negligible and, although giving no specific figures, he was of the opinion that the cost for single-phase equipment would not exceed that for third rail.

Eoghan's father was a very kind and generous person, always endeavouring to help those in need. He gave presents to the villagers from Killiney when they got married and when Patrick Doyle, one of the villagers who was a locomotive fireman on the DW&WR, got dismissed for telling a lie to screen his driver on the occasion of an accident in October 1906, he found him work. Murrough obtained a good character reference from Richard Cronin and arranged with Eoghan to employ him in the Power House at Formby. Regrettably, Patrick Doyle was to die of consumption in 1910. In November 1906 Murrough arranged employment at Formby for another young man from Killiney. James Lawlor, who was for 13 years a porter at Harcourt Street station in Dublin, was dismissed for no other reason than being absent from duty for a short spell when he had to urgently attend to his sister whose husband was in an Asylum.

When Eoghan and Frances had visited Killiney at the end of October, Frances shared a bit of exciting news with the O'Brien and Smythe families – she was expecting a baby in March 1907. Ellen went over to Formby to spend the last week of November at *Granagh* and during her stay she lunched with Mrs Aspinall who was delighted to hear of Frances' good news. Rather than travelling to Killiney for Christmas, Eoghan and Frances decided to stay at *Granagh*, where Murrough and Ellen joined them for the festive celebrations. Frances left no stone unturned to make it a most memorable occasion for everyone, and Ellen took the opportunity to urge her son to get a bigger house in time for the birth of the baby.

Silverdale, the detached house at Formby that Eoghan O'Brien purchased in February 1907. - [Courtesy Prof. O.F. Robinson]

On 19th January 1907 Eoghan wrote to his parents to say that he was in the process of negotiating for a property called *Silverdale*, a detached house in Formby standing on just over ⅓ acre. The deal was closed for £1,000 by 14th February when he was also able to tell his father that in order to meet the additional electrical power demand arising from the extensions to the electrified network and the increased levels of service a 4,000kW turbine driven alternator set was to be installed at the Power House in Formby. An additional combined rotary converter and battery sub-station was also commissioned at Aintree in 1907, and he confided that there was also talk of a further extension of the electrified network out to Ormskirk.

4,000kW turbo-alternator installed in the Power House at Formby in 1907 - [HOR-F-601 © NRM, York]

On Easter Day, 31st March 1907, Eoghan and Frances were blessed with the gift of a son, Brian Eoghan, their only child, who was born at *Silverdale* at 10.0 a.m. At the beginning of May, on their way to Switzerland for their annual holiday, Ellen and Murrough called in to see their grandson and they had hoped that the family would come over to them at Killiney in mid-July. However this was not to be, for in the first week of July correspondence commenced with George Hughes regarding William Arter, Eoghan's assistant on the Liverpool & Southport electrification, and another distinguished pupil of the Horwich school. Arter had got into trouble about patents, and Hughes had decided that it was no longer desirable to retain his services in his role at Formby. Hughes had spoken with Arter about the matter on the morning of 5th July; instructing him to immediately hand over to H.E. O'Brien any drawings or documents that might be the property of the L&YR. He had also indicated to Arter that he had arranged to find him some employment at Horwich Works should he wish to remain with the company, but Arter declined to accept the proposition and left the L&YR on 9th July.

Despite the delay to the start of their family holiday, Eoghan and Frances sent baby 'Brian Boru', and nurse Alice Kilbride on ahead to *Mount Eagle*, and they followed on 29th July. Eoghan had to return to Formby on 3rd August and it was eleven days before he was back in Killiney. On the day that he got back his father experienced abdominal pains, which Dr Henry Bewley considered might be an ulcer, a problem that periodically flared up to give Murrough trouble for the rest of his life. However, by November he was well enough to go over to Formby, and it was Frances who had to meet him at Liverpool as Eoghan had been called out to attend to an incident between Freshfield and Aintree. At 11.0 p.m. on 30th November a short circuit caused a rail to burn out and he was out on site until 4.0 a.m., and on 2nd December Murrough went with him to hear the explanation for the cable break that had caused the problem.

Experience of the first four years of operation of the electrified Liverpool – Southport service led to Eoghan preparing a paper entitled *Electric Traction on Urban and Inter-urban Steam Railways* that he presented to the Liverpool Engineering

Society on 22nd April 1908, and for which he won the Derby Gold Medal. In it he referred to the slow progress in relation to electric traction in England noting that, apart from the London Underground railways, only a short length of the Midland Railway at Heysham had been completed since the electric services on the NER and L&YR suburban lines had commenced. The LB&SCR scheme was still under construction, but no other English railway company had shown any inclination to adopt electric traction. He pointed out that, on the other hand, considerable progress had been made in America and on the Continent and quoted details of the most notable schemes.

Eoghan and Frances with Brian Eoghan at Mount Eagle in July 1908 - [Courtesy Prof. O.F. Robinson]

Brian Eoghan with his grandfather at Mount Eagle in July 1908. - [Courtesy Prof. O.F. Robinson]

Eoghan, Frances, baby Brian and Ellen relaxing at Mount Eagle in July 1908 - [M.J. O'Brien, Courtesy Prof. O.F. Robinson]

He stated that the problem of electrification of an existing steam railway was fundamentally a financial one; it was secondly a railway engineering problem, and lastly an electrical engineering problem. He stressed that this order of importance must be insisted upon owing to the tendency, on the part of electrical engineers, to neglect the financial and railway engineering sides of the question. He went on to state that, due to the large capital expenditure involved, either a sufficient decrease in operating costs must result in order to offset the increased fixed charges on capital, or the service must attract a growth in patronage to increase traffic receipts to cover both the additional operating costs of the increased service and the charge on capital.

An analysis of the performance of electric stock compared with steam hauled trains followed; H.E. O'Brien illustrating his points with comparative costs per train mile and expanding on the benefits of electric traction for suburban working. He compared the single-phase overhead AC and the direct current third rail systems and explained in detail the relative performance of AC and DC motor equipments. In analysing the capital and operating costs he concluded that there was little to choose between the two systems. However, he pointed out that the estimates he had presented did not allow for the alterations to structures and signalling that would be required for to install the overhead system; the cost of which would entirely depend on the physical configuration of the line.

H.E. O'Brien concluded that electrification could only be successfully adopted on existing railways if a large increase of traffic could be predicted consequent on a more rapid, more frequent, and cleaner service being delivered. He was also of the opinion that for suburban service the DC third rail system was the most suitable method of electrification, and that electrification of main lines for goods and passenger services was remotely probable. In the discussion that followed, Oliver Winder thought that the most important point which one could extract from the paper was that the choice of a particular system of electric traction had to be carefully made for the particular railway under consideration. A.D. Jones agreed that main line electrification was not likely to come to the front for some time, but that electrification was an excellent way of conducting suburban traffic through thickly populated districts.

The restrictive nature of the direct control stock led to the construction of the first multiple unit electric stock for the Lancashire coast lines during 1907-08. The order comprised twelve third class motor cars and six third class trailers. The motor cars were built with half-width motormen's compartments and gangways were provided at each end to enable the cars to run in any position in a train formation. The power bogies that were used on the original stock had not proved to be entirely satisfactory so a new improved design of motor bogie with a 9ft. wheelbase and leaf springs was adopted. The six motor cars built to Diagram 88 had no luggage space, but a luggage compartment with roller shutter doors was incorporated at one end of each of the six motor cars built to Diagram 89. The six third class trailer cars (Diagram 96) were built to a similar outline.

The year of 1908 was one in which the young family experienced health problems. During a visit to *Mount Eagle* in April, Frances consulted Mr Smyly, a specialist, who ordered her to hospital right away. She had a two-hour operation on 8th April at the Portobello Private Hospital in Dublin, and it was three weeks before she left hospital. Then, in early June, Eoghan was taken ill with pleurisy and there was, at first, a fear that he might have contracted tuberculosis. He was very run-down and had to take sick leave from 1st July but, of course, he was anxious to return to work as quickly as possible and his inclination was to get back in two months. George Hughes wrote to his father, to prevail upon him to take the full three months holiday "no matter whether he is improved in condition at the end of two months", as Hughes thought it highly desirable that he should not worry himself and come back to work too soon. George Hughes' intervention had the desired effect and it was not until 5th October that Eoghan returned to Formby.

In mid-October William Arter called on the O'Briens at *Mount Eagle*. By that time he was working in the USA and he was keen for Eoghan to join him over there and set up in a box manufacturing business. Arter and two friends, Collette and Roots, had through Roots obtained a licence to manufacture boxes using patent machinery and had formed the Bridgeport Wirebound Box Corporation in Connecticut. Collette and Roots were unable to raise the finance required, but Arter had borrowed £2,500 and wanted Eoghan to invest £2,000 or more. This would have meant giving up his job with the L&YR, losing of money on the sale of *Silverdale*, risking and probably losing what little capital he had, and becoming a slave to Arter. Murrough reckoned the whole scheme was suspicious, incomplete and indefinite; and thought it dishonest to ask others to invest in it, and so he advised Eoghan to steer clear of the proposal. Nothing more was ever heard of the scheme, but Arter subsequently became head of the railway department of the Allis Chalmers Company.

Towards the end of the year Eoghan was back at the I.C.E. to hear the paper on *The Single-phase Electrification of the Heysham, Morecambe and Lancaster Branch of the Midland Railway* presented on 9th November by James Dalziel and Josiah Sayers of the Midland Railway. In contributing to the discussion he stated that he wished to say a few words on the comparative weight and acceleration of trains worked by single-phase and direct current, without expressing any opinion as to their merits for particular types of service. H.E. O'Brien compared figures given by Messrs Dalziel and Sayers for the MR trains with those of the L&YR stock and demonstrated that the MR train required 9.2 kW/ton for an acceleration of 1.1 mile/h/s against 9.17 kW/ton for 1.25 mile/h/s of the L&YR train, noting that the L&YR train weighed 72.5 tons against 51.5 tons for the MR train. In reply to his comparison, the authors noted that the L&YR train had electrical equipment weighing 0.028 ton/hp, whereas the revised single-phase equipment on the Heysham line would weigh 0.029 ton/hp. They were of the opinion that "Mr O'Brien's train was a triumph of coach-building and not of electrical equipment," and wondered if it would stand up to the knocking about that the Heysham stock could withstand.

On Thursday 21st January 1909 yet another serious accident occurred on the L&YR

CHAPTER SIX Life with an Electric Railway

electrified section. On this occasion the 7.10 a.m. passenger train from Liverpool to Southport was standing at the Down fast line starting signal at Marsh Lane Junction when it was run into in the rear by the 7.20 a.m. passenger train from Liverpool to Hall Road. About five minutes later the 7.0 a.m. train from Southport to Liverpool came to a halt on the Up fast line opposite the two former trains due to traction power being cut off. The 7.10 and 7.20 a.m. trains were three-car units and were respectively carrying nine and three passengers, but the 7.0 a.m. from Southport was a four-car unit with seating for 292 passengers and was well loaded. The collision was not severe as the 7.20 train was proceeding cautiously, but arcing started a fire in its motor compartment, which spread to cars in the two other trains.

The collision occurred just beside the sub-station at Seaforth and the crews of both trains acted promptly in getting the few passengers out of each train and conducting them to a place of safety. When the 7.0 a.m. from Southport came to a stand opposite the fire the passengers were detrained and placed on an embankment clear of the rails. There were no injuries to passengers, but the motorman of the 7.20 a.m. train was pinned between the controller cabinet and the doorway to the luggage compartment and there was some delay in releasing him. Third class motor car No.3045, which was at the rear of the 7.10 a.m. train was badly damaged by fire, and the first class trailer next to it, No.429, also suffered collision and fire damage. On the 7.20 a.m. train, the greater part of the body of the leading third class motor car, No.3009, was burnt and No.427, the first class trailer next to it, was damaged by the collision and its body was partly burnt. The greater part of the bodies of first class trailer No.407 and third class motor car No.3050, of the 7.0 a.m. from Southport, were also badly damaged by fire.

Once more, it was Lieut-Col. Edward Druitt who conducted the formal inquiry for the BoT, on this occasion assisted by Mr A.P. Trotter.[54] In compiling his report for the Inspecting Officer, Trotter discussed the details with H.E. O'Brien, who informed him that an examination of the ruins of third class motor car, No.3009 showed no trace of fusion on any of the steelwork. Trotter concluded, therefore, that an arc had occurred between the positive and negative cables and that the fire started from that cause. Since the arc occurred on the cables before the circuit breakers they would not detect the excess current draw. H.E. O'Brien confirmed that it had been necessary to set the circuit breakers to trip at about 1,500A each in order to cater for the situation where only one or two of the eight collector shoes were in contact with the conductor rail and collecting the whole of the starting current for the train. From experiments that he conducted, which demonstrated that shoe fuses blew at 1,100A, it was clear that such fuses would blow if only one or two shoes were in contact with the conductor rail when a train started.

The chart of the automatic recording ammeter at the Hall Road battery sub-station

---

54 National Archives, PRO, MT1053/98 BoT Returns and Reports of Railway Accidents, 1909

showed that a sudden rush of about 2,700A passed at the moment of collision, but that no appreciable current flowed afterwards for about 3½ minutes until the 7.0 a.m. train started from Southport. About one minute later the short circuit caused by the driver placing the shorting bar on the line was evident. This all showed that the fire was not caused by any continued passage of high current; the question of the fuses and the circuit breakers could therefore be dismissed. Trotter concluded that the fire broke out gradually and that at no time were the passengers in any danger. The gradual growth of the fire was consistent with the ordinary spread on combustible materials, namely, wood and cable insulation. Lieut-Col. Druitt added that he did not think that the adoption of end doors in the carriages presented any extra risk from fire to the passengers.

On 15th February 1909 the section of the triangular junction between Hawkshead Junction and Meols Cop, and the easterly portion from Meols Cop to Roe Lane Junction were opened for electric trains. Thereafter trains ran to and from Crossens via Meols Cop, reversing at the latter station, instead of via the direct line, although some peak-hour trains continued to use the direct route. In 1912 construction was commenced on a new electric stock maintenance workshop inside the triangle of electrified lines at Meols Cop. The old depot at Southport had eventually proved inadequate and the spacious new workshop, which was brought into use in 1913, was a great improvement. General overhauls and heavy repairs were, however, still dealt with in the main workshops at Newton Heath.

On 8th April 1909 Eoghan happened to be in Dublin when John Hall Rider read his paper on *The Electrical System of the London County Council Tramways*. In joining in the general chorus of congratulations to the author he said that he had only one or two criticisms to make; one being that he felt that the 10% moisture content in the coal specification was on the high side. The other point he raised was that the LCC Tramways seemed to have 25% spare capacity in their boilers, an estimate that was confirmed by Mr Rider. H.E. O'Brien felt that this was excessive bearing in mind the extent to which water-tube boilers could be forced if required. In commenting on superheat, he wondered at what steam temperature the author was working his Corliss engines, as at Formby he had found that 510°F seemed to be the extreme limit for the ordinary Corliss valve gear. He was also interested in what the author said about cross currents between generators. At Formby he had put ammeters in the neutral and not found any cross currents between the four 1,500kW alternators when running in parallel. He had, however, found a considerable cross current, of about 50A when working at 7,500V, once the 4,000kW turbine generator was put on in parallel with the reciprocating plant.

In May, Eoghan was notified that he had been selected to fill the position of Assistant Carriage & Wagon Superintendent at Newton Heath, Manchester, at an annual salary of £500 per annum from 12th June 1909. The vacancy had arisen as a result of the resignation of Oliver Winder, who left the L&YR to become Manager of the Patent Shaft & Axle Company Ltd. at Wednesbury. J.P. Crouch was promoted from Newton Heath to fill the position of Assistant Chief Mechanical Engineer and Works Manager, Horwich, and C.H. Montgomery replaced O'Brien at Formby. Despite the move, H.E. O'Brien was still considered to be the L&YR 'Engineer for Electric Traction' and so the title adopted for Montgomery's position was Resident Electrical Engineer, Formby.

The I.Mech.E. Summer Meeting in 1909 involved a visit on 29th July to the electric rolling stock repair shop at Southport, the Power House at Formby and Horwich Works. Although H.E. O'Brien had moved to Newton Heath a few weeks earlier, he had been responsible for the arrangements for the visit to Southport and Formby and was on hand on the day of the visit. For the journey over the electrified railway to Southport and Formby and return, the L&YR provided a special train comprising five first class cars, which left Exchange station, Liverpool, at 9.25 a.m. and ran the 18½ miles non-stop to Southport in 30 minutes. The special left Southport for Formby at 10.45 a.m. and, following the visit to the Power House, proceeded to Liverpool. After luncheon at the hotel at Exchange station, at which Sir George Armytage presided, supported by the President of the Institution, J.A.F. Aspinall, and officials of the L&YR; the party visited the locomotive works at Horwich in the afternoon. Following the tour of the Horwich works, tea was served at 4.0 p.m. in the Mechanics Institute.

Management of Workshops

As a result of a serious fire that had destroyed the carriage shops at Miles Platting on 27th April 1873, the L&YR had decided to build new facilities at Newton Heath, about two miles northeast of Manchester Victoria station. Construction had commenced in October 1874 and was approaching completion when Frederick Attock took up duty as Carriage & Wagon Superintendent on 1st February 1877. Although wagon building had continued at Miles Platting following the opening of the Newton Heath workshops, the facilities at the former were far from adequate for the purpose, and so Attock submitted a report in January 1889 regarding provision of a wagon shop at Newton Heath, which was constructed during 1889-90. The purchase of additional land at Newton Heath was completed in January 1895, and new carriage shops were erected on the enlarged site and opened in 1898. From that time onwards, carriage building was concentrated in the new shops, wagons being largely dealt with in the older buildings.

Eoghan O'Brien's promotion included the provision of a company house close to the works at Newton Heath. He had to pay a nominal rent of £27 per annum for the benefit, but the L&YR agreed to pay the rates and taxes and keep it in full repair. In addition, a Brougham was kept at the house and a man was on hand to drive him the 2½ miles to Hunts Bank, Manchester, when required. Although the move meant giving up their own house, and going to live in Manchester might have proved trying for the family, *Ely House* turned out to be located in a favourable position on high ground with a garden behind it. Commenting to Ellen on his move, J.A.F. Aspinall said "He is no longer on a siding, but on the mainline for promotion," and indeed that was the case as he had become the No.3 mechanical engineer on the L&YR. George Hughes had hoped to visit Newton Heath during the first week of his appointment, but in the event was not able to do so. However, Hughes felt that his inability to visit on that occasion was not a bad thing as it would give O'Brien the opportunity of "shaking down into some of the detail", and he also assured him that F.E. Gobey, who had taken over from Montgomery as Works Manager, and all of the staff at Newton Heath would be in a position to give him every assistance he needed.

Eoghan and Frances took a three-week holiday at Killiney from 18th September; baby Brian and nurse Kilbride having gone ahead of them a week earlier. During their stay, Eoghan bought an automobile which had been owned by his brother-in-law, Rupert Caesar Smythe. This was the beginning of Eoghan's life-long love of motoring, and his mother was delighted when he took her for a drive around Coliemore, Bullock, Castle Park, Sallynoggin, Rochestown Avenue and back to *Mount Eagle*. The family was back at *Ely House* on 11th October, but Eoghan's period in charge at Newton Heath was to be less than nine months. The total number

CHAPTER SEVEN **Management of Workshops**

Eoghan's first automobile, which he purchased in October 1909 for £30 from Rupert Caesar Smythe, his brother-in-law. Murrough noted that Ellen was 'infected' and wanted one too! - [Courtesy Prof. O.F. Robinson]

of bogie passenger carriages constructed during his tenure was 213 (see Appendix Three), and the carriage portion of Railmotor No.15 and at least six horseboxes were also completed during 1909. The production in the wagon shops continued unabated; 1,371 vehicles being completed in 1909, and 1,250 in 1910. Although it is not possible to ascertain the precise number of freight vehicles constructed whilst he was in charge, it is fair to assume, on a pro-rata basis, that at least 1,000 were completed at Newton Heath during O'Brien's time (see Appendix Four).

The main carriage shop at Newton Heath - [HOR-F-4079 © NRM, York]

A view inside the lifting shop at Newton Heath - [HOR-F-4068 © NRM, York]

On 1st November, C.W. Bayley gave notice of his intention to retire on 1st March 1910 and the L&YR Board promptly decided to promote J.P. Crouch to Passenger Superintendent, and appoint H.E. O'Brien in his place as Assistant Chief Mechanical Engineer and Works Manager, Horwich, on a salary of £750 p.a. The role of Assistant Carriage & Wagon Superintendent went to F.E. Gobey, and R.G. McLaughlin was appointed Works Manager at Newton Heath. In telling his parents of his promotion, Eoghan wrote from *Ely House*:

> Some astonishing news for you; owing to changes consequent on the retirement of the Chief Traffic Manager, I become Assistant Mechanical Engineer at £750 p.a. and we have to move to Horwich. The appointment is to date from 1st March, so that we shall be here until the end of February. I am glad in some ways, but sorry not to have had a longer experience here, and sorry too for those whom my promotion will disappoint. We shall have a nice house at Horwich and be more in the country.

Just before Christmas 1909, Mrs Aspinall wrote to Frances, as follows:

> I never wrote to congratulate Eoghan on his promotion, which has indeed been rapid, and a reward of hard work and loyalty, which few men find so young. I think that you will find Horwich a few degrees better than Newton Heath as a place of residence, but it is a bleak spot. I was never over the house *Overdale*, so I have no idea whether the accommodation is better than *Ely House*, or whether the decoration will be to your taste. Having been done up so recently I suppose you will have to make the best of things as they are!
>
> If you are not going anywhere else, will you come to us for Christmas day; bring Nurse and Brian with you if you don't think the weather more inducive to staying by your own fireside, but at any rate you will know that I have been thinking about you.
>
> I don't know if Mr Aspinall acknowledged Eoghan's letter, but I do know that he much appreciated it. It is a pity more people do not take the trouble to express thanks for what one does for them; it costs so little and means so much to those in authority who, after all, are very human.
>
> With love and all good wishes from your friend,
>
> Gertrude Aspinall

However, one issue that required the attention of H.E. O'Brien before he left Newton Heath was the appointment of a Chief Accountant at the Carriage Works

CHAPTER SEVEN **Management of Workshops**

to replace J.T. Grime who had died. The job was given to S. Smith, and P. Pilling was appointed the Assistant Accountant.

During his time at Newton Heath, Eoghan had been working on the design of an improved buffer for use on railway vehicles and had submitted a Patent application on 23rd August 1909. Up to that time self-contained buffers had only room for a spring action of about three inches owing to the spring being placed entirely on the inside of the buffer casing. O'Brien's invention provided increased room for spring action of up to 5in., thus reducing the longitudinal stresses and strains imparted in the vehicle underframes by buffering action. The Patent[55] was accepted on 11th August 1910, but it was subsequently found that, with the main spring not entirely encased, the partial casing did not form a positive stop for the buffer plunger. The problem was overcome by an improvement that he subsequently developed with Walter Gatwood, an engineering colleague. This involved forming a ring on the last coil of the volute spring to enable the spring to be secured to the buffer base plate by means of an outer flanged ring. A patent application for this improvement was submitted on 11th March 1911 and the Patent[56] was accepted on 8th February 1912.

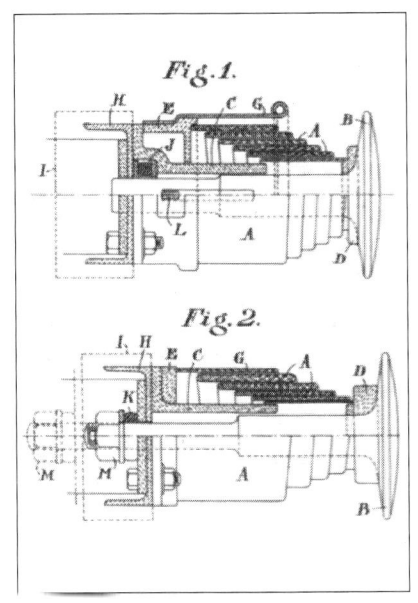

Drawing of H.E. O'Brien's Buffer Patent of 1909

Official L&YR photograph of Eoghan O'Brien issued on the occasion of his appointment as Assistant Carriage & Wagon Superintendent in June 1909, and subsequently re-issued on his promotion to Assistant Mechanical Engineer and Works Manager (Horwich) on 1st March 1910 - [Reproduced from Railway Gazette April 15, 1910, by permission of the Editor]

---

55 Improvements in Buffering Devices (British Patent No.19,362 of 1909)
56 Improvements in Self-contained Spring Buffers (British Patent No.6,136 of 1911)

One of the first events that Eoghan was involved with at Horwich was the concert and prize distribution in connection with the evening classes held at Chorley New Road Council School in May 1910. In a short speech before distributing the prizes he stated that education seemed to him "the most important thing in the whole world, for upon it depended the men and women of tomorrow". He was afraid, though; that ratepayers might not agree with him, as in England insufficient money was spent on education. He pointed out that the danger from Germany and Switzerland lay in the competitive advantage of their education, which was a greater threat than that of their armies, adding that in Germany and Switzerland nearly twice as much was spent per head on education as in England. He said that when visiting those countries he had been struck by the number of people who could speak three or four languages and with the extent of useful knowledge that they possessed.

The 8th session of the International Railway Congress was held at Berne, Switzerland, from 4th – 16th July 1910, and H.E. O'Brien was due to attend as a representative of the L&YR. Amongst the other delegates were some of his close associates of long standing: his General Manager at the L&YR, J.A.F. Aspinall; his former mentor, Richard Cronin of the D&SER; and a former colleague of both Aspinall and Cronin from their Inchicore days; Robert Coey, Locomotive Superintendent of the GS&WR. However, Eoghan suffered a slight recurrence of the complaint that had resulted in him taking three months sick leave during the summer of 1908 and in the end only managed to attend for one day. In order to recuperate, he stayed on with Frances and her sister Barbara (Bay) in Switzerland, initially residing at the *Hotel du Chateau*, Ouchy-Lausanne.

At the same time J.A.F. Aspinall was also staying in Ouchy-Lausanne at the *Beau Rivage* hotel, as he also required an opportunity to recover from the stresses of his work. On one occasion they all went to Geneva together, and Frances recorded that Mr Aspinall was charming and that the rest had done him a lot of good, although he found it hard to do nothing! While in Switzerland, Eoghan went to Berne to get his railway passes stamped so that he and Frances could travel for free on the trains and steamers anywhere they liked in Switzerland. Although he was slowly recovering, if Eoghan attempted anything too strenuous, like going on a train journey or walking about too much, his temperature would rise. Aspinall and Hughes both insisted that he must take a good holiday, not worry too much, and rest until he got strong again.

Before leaving Ouchy-Lausanne for a week at Saas-Fée, they took a trip to Chamonix, during which Eoghan took some notes on the electric railway from Martigny, which he found most interesting. Bay also seemed to enjoy her time in Switzerland, taking an interest in the flowers, geology and the history of the region. While Eoghan and Frances were in Switzerland, young Brian stayed at *Mount Eagle*, with his nurse, Alice Kilbride. However, after 3½ years unremitting care and kindness she decided to give up her position when the family returned to *Overdale* in September. Brian's grandparents valued her service so highly that Murrough gave her a cheque as a token of their appreciation, but Ellen noted that, "no money could buy what she had given".

An extension to the electrified network, from Aintree to Maghull, had been completed whilst H.E. O'Brien was at Newton Heath, and opened on 1st October 1909. Despite his new role as Works Manager, he was still the acknowledged electric traction expert on the L&YR and was regularly consulted by George Hughes on any matters concerning electrification. On his return to Horwich in September 1910, a memo from Hughes regarding a further stage in the electrification programme, on the mainline northwards from Maghull to Ormskirk, was awaiting his attention. In it Hughes outlined the difficulties that had been encountered with landowners in regard to the erection of an overhead high-tension feeder line between Formby and Ormskirk and advised that Aspinall had made up his mind to abandon that part of the scheme. It was decided to consider its substitution with an overhead high-tension cable from Aintree to Aughton Park, but Hughes felt that, if a return on the investment in the Ormskirk extension was to be obtained almost immediately, redundant cable between Hall Road and Formby could be recovered and re-used to reduce costs. Of the nine miles of cable available, 1½ miles was, however, required for the Waterloo widening.

L&YR Chief Mechanical Engineer's departmental offices at Horwich - [HOR-F-466 © NRM, York]

Hughes considered that the recovered cable might be included as part and parcel of the scheme for connecting the Power House at Formby to Ormskirk via Hall Road, Seaforth and Aintree. He therefore asked H.E. O'Brien for his opinion on whether it would be commercially sound to strip the recovered cable of its old armour and send it back to the supplier (Messrs Glover), or to some other firm, to be re-armoured. However, Hughes thought that it would not be necessary, and that if the cable were properly placed in raised boxing and covered with bitumen, there should be no further trouble from short circuits, but he required him to investigate the matter and confirm the appropriate action. A further point that would arise in connection with the proposed use of the recovered cable for the feeder was the need to introduce step-up transformers to maintain the voltage, which would result in additional expenditure, maintenance and interest charges, and Hughes requested O'Brien to take such factors into consideration when undertaking his assessment.

The Electric Motor Shop at Horwich with Foreman T. Clayton in the centre background - [HOR-F-2785 © NRM, York]

Regardless of whether the recovered cable was used or a new overhead high-tension line was constructed, Hughes confirmed that the sub-station at Aughton Park should still be constructed to a design suitable for accommodating the maximum amount of machinery envisaged for future requirements, and left in such a manner that it could be easily extended when required. Hughes also pointed out that the feeder line should be run on the east side of the railway between Aintree and Ormskirk, as the proposed widening scheme was to go on the opposite side. Because the turbine driven alternator at Formby was giving trouble Hughes also asked H.E. O'Brien to have some of his staff go into the question of installing exhaust steam turbines in conjunction with at least two of the reciprocating engines at Formby Power House. The extension from Maghull was opened to Town Green on 3rd July 1911, but it was to be 1st April 1913 before services commenced on the Town Green to Ormskirk section.

Another subject that was raised in the same memo was a scheme for the construction of an experimental electric locomotive that could be used to gain some experience on passenger service and in shunting at Aintree. During O'Brien's absence, Hughes had instructed draughtsman Wadsworth to make himself familiar with the situation

CHAPTER SEVEN **Management of Workshops**

in regard to the operation of interchange of traffic between Fazakerly and Aintree sorting sidings, and also to arrange for tests to be carried out with a Liverpool & Southport motor car to ascertain its maximum haulage capacity on shunting duties. The schemes for the locomotive envisaged a cheap experiment, and proposals had been drawn up on the basis of utilising the frames and wheels of a radial tank locomotive, a standard goods locomotive, or an outside cylindered six-wheeled coupled shunting locomotive. Standard L&YR electrical equipment was to be employed, and the schemes as first developed considered power transmission by means of chains manufactured by Hans Renold Ltd of Manchester. Hughes had shown the preliminary proposals to Aspinall, who was favourably impressed with the radial tank scheme, but had requested the various options to be further developed before he made a definitive decision.

L&YR experimental electric locomotive 'under the wire' at Aintree in 1912; the collecting shoes for operating on the third rail can also be seen
- [HOR-F-1066 © NRM, York]

View of the Niesen Funicular Railway used by H.E. O'Brien in Continental Engineering, his Presidential Address to Engineering & Scientific Club of the Horwich Railway Mechanics Institute, November 1910 - [H.E. O'Brien]

For the purpose of operating the locomotive in the Aintree yard, a number of the lines were electrified at 600V DC. An overhead catenary system was adopted as it was felt that live third-rail in the yard would pose too great a hazard. The final design chosen for the electric locomotive was indeed based on the frames and wheels of a standard 2-4-2 radial tank, and four standard motor car traction motors were geared in pairs to two separate jack-shafts. Power was transmitted by means of cranks on the jack-shafts via outside connecting rods to the driving wheels. The body was of steeple-cab design, and current collection was via pantographs mounted just below roof level on each cab end, as well as via collector shoes for working in third rail territory which were live when current was being collected via the pantographs!

L&YR electric locomotive No.1 went into service in early 1912 and was used for haulage of coal between Aintree sorting sidings and North Mersey sidings as well as on shunting duties at Aintree yard. For the trip to North Mersey sidings it worked under the overhead line from Aintree yard for the short distance to Sefton Junction where it then went onto the third rail previously laid for passenger services. The locomotive was temporarily withdrawn during the war years, but reappeared in 1919 on passenger duties between Southport and Crossens hauling two of the lightweight cars built for working over the LOR lines. This running was considered an experiment and did not last long and the locomotive was taken out of service and stored at Meols Cop electric car depot. It was officially withdrawn in 1920 and the overhead line equipment in Aintree yard was dispensed with around the same time.

Reverting to 1910, Frances was suddenly taken seriously ill in November and Sir William Sinclair decided that she must have an ovarian operation, which took place on 23rd at a private nursing home in Bolton. Her sister Bay, who had joined the family at Horwich in April and had been with them during their month in Switzerland, took charge of affairs at *Overdale*. It was well that she was there, as Eoghan was very busy as usual in the Works with new problems turning up that continually required his attention, and details of a very complicated organisation to be grasped as well as much technical work to be dealt with. As a result of all that was happening, the family did not travel to Killiney for Christmas.

H.E. O'Brien had been elected Chairman of the Horwich Railway Mechanics Institute Engineering & Scientific Club and on 25th November, in his address for the 1910/11 session, he spoke on the subject of *Continental Engineering*. He illustrated his talk with a dozen of his own photographs including views of the Niesen Funicular Railway, the Mount Pilatus Rack Railway, a rack locomotive for the Brunig Railway, an electric locomotive for the Bern-Lötschberg-Simplon Railway and a superheated locomotive for the Gotthard Railway. In the following February he made his first visit to Belgian State Railways in Brussels and then went on to Messrs Siemens in Berlin.

Eoghan was in attendance at the I.C.E. on 14th March 1911 for Philip Dawson's paper *Electrification of a portion of the Suburban System of the London Brighton & South Coast Railway*. The LB&SCR had successfully introduced a 6,700V AC overhead system on 1st December 1909 on the South London Line between London Bridge and Victoria; and an extension to the system from Battersea Park Junction to Crystal Palace, and onwards to Norwood Junction, was nearing completion. In contributing to the discussion on the paper, H.E. O'Brien remarked that the battle between advocates of the direct current and single-phase alternating current systems had often been mentioned. He pointed out that, although he was probably classed among those who were advocates of the direct current system, he was not averse to the single-phase system, but did not see it as a universal solution. He then asked the author if the single-phase system was to be preferred to the direct current system for suburban operations under all circumstances, adding that the additional capital cost of £4,000 or £5,000 per mile for the single-phase system would be impossible to bear without any adequate return even on mainline schemes. He felt that no one had explained what return there would be for such expenditure; was it to be in repairs, wages of the firemen or drivers, or in coal?

In reply, the author admitted that for tube lines in London the direct current system was satisfactory, but that for the electrification of any portion of main line railways he felt that it was practically unthinkable. He added that this was the same conclusion that all the principal railway engineers of Europe had arrived at. Further technical discussions ensued on the relative merits of the traction motors employed in the respective systems, H.E. O'Brien drawing attention to the great inferiority of the motor used in single-phase traction, and insisting that it was the weak point in the application of the single-phase system to suburban traction. Suffice to say that, in the long term, the southern suburban lines out of London were eventually all electrified on the third rail direct current system.[57]

On 8th March Ellen fell off a set of steps in the linen closet at *Mount Eagle* and broke her fibula. Dr Pim set the broken bone, and an excellent, attentive and cheerful resident nurse, Delia McDonagh, was brought in for six weeks to attend to her needs. She was sufficiently recovered by 3rd May to travel by the 9.15 a.m. train from Kingsbridge to Blarney where she was to stay at the Hydro for further recuperation. However, she had only been gone three days when Murrough began to feel unwell again, but still he managed to travel to Blarney on 12th May. While there his condition deteriorated and he was in such pain that on 17th the local doctor thought he would not last through the night and injected him with morphine and strychnine. The next day Dr John Dundon came up from Cork to examine Murrough and arranged for him to be admitted to the Mercy Hospital in Cork. He lay between life and death for nearly five days and Eoghan, who had come over immediately, brought Ellen down from Blarney to the *Imperial* hotel in Cork to be near Murrough.

By 26th May his father had recovered sufficiently for Eoghan to arrange for first class seats in the Up Mail train from Cork at 3.30 p.m. The train reached Dublin on time at 7.25 p.m., but instead of staying on board to run around to the pier at Kingstown, a car met them at Kingsbridge station to take them straight to *Mount Eagle*. Once again Delia McDonagh was on hand, this time assisted by nurse Margaret Dunphy. Murrough began to recover, but then got an attack of pneumonia and became very ill, and for three weeks he was once again at death's door. Eoghan and Frances returned to Killiney and stayed at *Court-na-farraga* until 27th July, visiting his father daily until he had turned the corner. A surprise visit by their old friend E.L. Vaughan, who had dinner at *Mount Eagle* on 23rd August, was an event

---

57 The Southern Railway decided in August 1926 to abandon the LB&SCR single-phase system; the last train to operate on the old system ran from Victoria to Coulsdon North on 22nd September 1929.

that brought some cheer after over five months of ill health.

Meanwhile, back at Horwich, life was as busy as ever for Eoghan. Although he was regularly consulted on matters pertaining to electrification projects, his principal task was the management of the Locomotive Works. When he took over in March 1910 work had just commenced on the construction of a batch of ten saturated 2-4-2 radial tank locomotives equipped with Belpaire fireboxes. These were quickly followed by twenty 0-4-0 saddle tank locomotives, which were completed by July 1910, when work started on twenty 0-8-0 goods locomotives with large saturated boilers. In March 1911 a further batch of twenty radial tank locomotives were put in hand, but for that batch superheated boilers with Belpaire fireboxes were used. Seventeen had been completed by July, but at that point work at Horwich came to a halt.

In the summer of 1911 the workforce at Horwich aired their grievances in relation to certain aspects of work practices at the workshops. Although management had refused to recognise the Trade Union officials, George Hughes and H.E. O'Brien met with the Joint Committee of the unions in July. However, they declined to abolish the system of 'checking-in' at the Works' gates and also the system of fixing standard times for jobs. As a result of management's refusal to accept the men's grievances, the workers downed tools at 11.00 a.m. on Thursday 3rd August and sat at their work places until the lunchtime hooter sounded. Management did not close the gates and the men were able to return to their work places after lunch. In the mid-afternoon an official notice was posted at the main gate advising that the men who had been idle could signify their intention to work by picking up their checks as usual on Friday morning. The Joint Committee construed that action to be tantamount to a lock out, and the decision was taken to continue the strike.

On 9th August Hughes wrote to O'Brien advising that the men who had taken strike action would be paid for any time that they had actually worked in the week up to the time of stoppage. It was arranged for this to be done whether the men were connected with the conciliation scheme or outside of it, and some 4,000 men and boys were paid a full week's wage on 11th August. There was little animosity shown towards the top management, George Hughes and H.E. O'Brien being held in high esteem for having treated the Joint Committee members with courtesy during their various meetings. With the exception of a small demonstration by some women against one official, the friendly relations between the salaried staff and the employees remained unaltered. A demonstration of that good will was shown in the arrangements for one of the workers meetings. Young Brian Eoghan O'Brien had been taken seriously ill – with suspected poisoning – and, in order that the sick child and his family should not be disturbed by the noise of a crowd passing *Overdale*, pickets were stationed nearby to ensure that those going to and from the meeting would use an alternative route.

Management and supervisory staff took over many operations in the Workshops; the Assistant Works Manager, G.N. Shawcross, taking on duties such as shunting, driving and firing of locomotives, and firing the Lancashire Boilers that ran the high-speed engines and pumps in the Power House. H.E. O'Brien drove the locomotive with Hughes' coupé attached from Blackrod back to the Works with the stationmaster from Horwich acting as pointsman. The use of what the striking men considered 'scab labour' drew an angry reaction and police were quartered at some officers' residences for a fortnight in September in case mob violence got out of hand. H.E. O'Brien was certainly not anti-union, his politics being middle of the road, but he disliked all forms of extremism. In order to avoid his off-duty presence in Horwich provoking any additional antagonism, he took up residence outside of the town during September. Together with the foremen on duty, he would arrive at Horwich station by train and walk down the line to the Works accompanied by policemen, keeping out of sight of the pickets at the gate. On a number of occasions the train was stopped just short of the station so that they could easily alight adjacent to the Works.

Towards the end of the month a meeting with the L&YR Board and management was convened at the Works by G.C. Cummins of the BoT. Terms of settlement were drafted and submitted to the union officials for their consideration, but they were unyielding. Cummins chaired a further meeting on 30th September, at which union officials were in attendance, the L&YR being represented by George Hughes and H.E. O'Brien. The meeting lasted from 10.00 a.m. to 6.30 p.m. with a short break for lunch, and an agreement was hammered out on the basis that: "If the management promise that the labourers' question shall, upon resumption of work, have immediate consideration and, if not satisfactorily arranged within two weeks of such resumption, the matter to be arbitrated upon within another two weeks, then the labourers would agree to allow their question to stand over until a resumption of work and would then send in their demands on the day work was resumed."

It was a fine morning when work resumed at Horwich on 6th October and, with most of the attention on the men returning to work, H.E. O'Brien "slipped into the Works almost unobserved a little after nine o'clock".[58] On that day the men submitted a memorandum requesting consideration of a minimum wage of 20s per week for all adult labourers and an advance to all skilled labourers receiving 24s per week and under. The L&YR declined to accede to the request and on 28th November 1911 the men, acting under the agreement of 30th September, applied to the BoT to appoint an Arbitrator to determine the issues that had been raised. The BoT appointed James Valentine Austin, Judge of County Courts, to be the Arbitrator and hear and determine the matter in accordance with the Conciliation Act, 1896. The Arbitrator awarded and determined that:

---

58 *Horwich & Westhoughton Journal*, 7th October 1911.

1. The wages of all the workmen (including stores labourers and issuers of material) who are under 21 years of age and over, and whose present rating is under 20s per week, be increased to the day work rate of 20s per week.

2. The wages of all the workmen (including stores labourers and issuers of material) whose present rating is from 20s to 24s per week, be increased by 1s per week.

3. The workmen failed to satisfy the Arbitrator that it was either necessary or desirable that he should make any order, or give any direction with respect to the piecework rates, which are from time to time fixed by agreement between the Company and the men.

The Horwich strike had seriously interrupted the programme of locomotive overhauls and on 9th November 1911 O'Brien spoke with J.A.F. Aspinall about the question of obtaining boilers from outside firms. Although he did not find favour with the idea, H.E. O'Brien recognised that the nine-week strike had resulted in a failure to produce 18 boilers, which should otherwise have been done. It was agreed that the need to purchase boilers would be dictated by the boiler situation rather than by price or delivery and that, when the decision was arrived at to place orders, the L&YR would supply the boilerplates and that the boilers should be supplied without tubes. This was the same condition in which boilers normally passed from the Boiler Shop to the Erecting Shop at Horwich. Although such drastic measures had to be taken in order to catch up with locomotive overhauls, three weeks later Hughes was expressing his concern to H.E. O'Brien that the cost of materials for engine repairs that had been ordered from outside suppliers were in aggregate 43% higher than the Horwich prices.

Locomotive building at Horwich recommenced after the strike, but progress up to the end of 1911 was slow; only the three outstanding radial tank locomotives and two new railmotors being completed by the end of the year. The first of a batch of twenty 0-6-0 goods locomotives with superheated boilers and Belpaire fireboxes was turned out in March 1912, the order being completed in October. From that point onwards, until the outbreak of War, new work at Horwich was concentrated on the construction of 0-8-0 goods locomotives with large superheated boilers and Belpaire fireboxes. Three lots totalling fifty locomotives were completed by December 1914.

Meanwhile, Murrough's health continued to give cause for great concern, and on 15th November he consulted Dr Joseph O'Carroll of Merrion Square, Dublin, who put him on a special diet. He was back again with Dr O'Carroll and Dr Gordon in January 1912 and they advised that there was no need for an operation, but that Murrough should take complete rest. Then on 1st February Dr Gordon advised Ellen

CHAPTER SEVEN **Management of Workshops**

that she had to have an operation at once and she was admitted to the Portobello hospital where she was operated on two days later. Murrough's sister, Geraldine Wilson, went over to stay at *Mount Eagle* and visited Ellen every day. She left when Eoghan and Frances arrived on 23rd February, but although all of his parents' health problems were a great concern to them, Eoghan had still to undertake his duties for the L&YR.

Inside the Locomotive Erecting Shop at Horwich - [HOR-F-2823 © NRM, York]

The year 1912 saw further management changes in the Chief Mechanical Engineer's department. J.P. Crouch had earlier resigned his position as Passenger Superintendent on 31st December 1910 to take an appointment as Chief Mechanical Engineer of the Central Argentine Railway. Two years later C.H. Montgomery followed him there as Works Manager, Rosario, his position as Resident Electrical Engineer at Formby being filled by Harold Creagh from 1st May 1912. A month earlier A.D. Jones had left the L&YR on his appointment as Locomotive Running Superintendent on the South Eastern & Chatam Railway. F.W. Attock was appointed to succeed him from 1st April with H. Housley as his assistant in the Outdoor Locomotive department. Zechariah Tetlow, the long-standing Chief Locomotive Draughtsman, retired on 1st December and was succeeded by J.R. Billington. He had been in charge of the Gas department since taking over from G.H. Roberts, who had left the L&YR to join the Royal Armaments factories at the end of May 1904.

In February 1912 Messrs Dick, Kerr & Co. submitted a proposition to the L&YR to electrify the Holcombe Brook branch with a 2,500V DC system at their expense.

The 3¾-mile single track branch had one crossing loop at Woolfold; the ruling gradient was a stiff 1 in 40; Holcombe Brook being 525 ft above sea level. Dick, Kerr & Co. proposed this experimental electrification as they wished to gain experience of high-voltage DC working in preparation for tendering for a contract in Brazil. An experimental line was set up at their Preston Works and in order to facilitate trials with high-voltage traction motors the L&YR supplied a pair of Liverpool & Southport bogies without motors. Consideration was initially given to selecting some open third class carriages for conversion, but in the end it was decided to build some new stock, two motor cars (Diagram 133) and two trailers (Diagram 134). The actual operating voltage adopted was 3,500V DC with traction current supplied via overhead line rather than third rail. Power was obtained from the Lancashire Electric Power Co. and fed to the overhead system from a sub-station at Holcombe Brook and electric services on the branch commenced in July 1913.

Holcombe Brook branch - 3,500V DC trial, 1913 - [M. Blakemore collection]

A two-car experimental set comprised of Motor Car (Dia 133) and Trailer (Dia 134) at Holcombe Brook station [HOR-F-1262 © NRM, York]

F.W. Carter read his paper on *The Mechanics of Electric Train Movement* to the Local Section of the I.E.E. at Manchester on 16th April 1912. H.E. O'Brien was among those in attendance and contributed to the discussion. He considered that the question of train resistance was of great importance for the larger electrification schemes, which he felt were bound to take place in the future, although it was not of very great importance on suburban work. He pointed out that the resistance curves presented in J.A.F. Aspinall's Presidential Address to the I.Mech.E. showed lower train resistance at the higher speeds than those presented by the author. In regard to satisfactory methods for obtaining train resistance figures, H.E. O'Brien stated that delicate and accurate instruments were required and he referred to the Doyen's Inertia Ergometer used on the Belgian State Railways dynamometer car, and the author endorsed all that H.E. O'Brien said in relation to reliable train resistance figures for electric trains.

The Traffic Committee of the L&YR recommended expenditure of £19,715 for the electrification of the Town Green to Ormskirk section to the Board on 12th June 1912. Writing to H.E. O'Brien on 2nd July, Hughes advised that Aspinall had agreed to the route for the cable and overhead high-tension line from Birkdale to Ormskirk. Hughes then pointed out that consideration would have to be given to the lead time for the various pieces of plant so that the overhead high-tension line and the third rails would be in position in time for the delivery of the 900kW rotary converter for the Ormskirk sub-station. Work progressed rapidly and the extension to Ormskirk was ready in time for the commencement of electric traction operation on 1st April 1913.

In the meantime, Eoghan's extra-curricular activities continued. He was among the guests at a function to celebrate the 25th Anniversary of the Mechanics Institute, which was held at Horwich in September 1912. Aspinall had invited all of his past pupils to attend and amongst those who gathered with L&YR officers were many who had moved on to careers outside of the L&YR, including J.P. Crouch, H. Fowler, H.N. Gresley, A.D. Jones, G.H. Roberts, and O. Winder. In his Chairman's address to the Horwich Railway Mechanics Engineering & Scientific Club on 18th November 1912, H.E. O'Brien spoke on the subject of *More Continental Engineering*. Once again he illustrated his talk with some of his own photographs including views of the Simplon Tunnel, superheated passenger and goods locomotives for the Baden State Railways and a Sulzer Uniflow steam engine.

The surviving two-thirds of a photograph believed to have been taken at a gathering to celebrate the 25th Anniversary of the Horwich Mechanics Institute. Those persons identifiable include (seated l-r) H.N. Gresley, J.A.F. Aspinall, J.P. Crouch, G. Hughes, O. Winder, J. Tatlow, H. Fowler, H.E. O'Brien; (middle row l-r) J.H. Haigh, J.P. Hamer, R.G. McLauglin, J.R. Billington, F.S. Barnes, F.E. Gobey, D. Gibson, G.N. Shawcross, and Dr J.H. Jackson.

F.W. Attock, H. Creagh, H. Housley, A.D. Jones, A. Lund, R.E.L. Maunsell, C.H. Montgomery, and G.H. Roberts were probably amongst those included to the left in the missing part of the photograph. - [Provenance unknown]

Eoghan's aunt Kate Toulmin died on 16th August, at Hatfield Peverel, Essex, and

Murrough had been well enough to attend his sister's funeral. However, he again started to get bouts of abdominal pain in September and on examination by Dr Pim a duodenal ulcer was diagnosed, but he felt that there was no need for an operation. This was better news for Eoghan, who had made arrangements to visit the depot at Scheveningen for the electrified Rotterdam – Den Haag line in Holland on his way to Brussels on 1st October. His trip to Brussels was for the purpose of a further visit to the Belgian State Railways workshops and for discussions with Jean Baptiste Flamme (1847-1920), the first engineer to successfully fit Schmidt's firetube superheater in a steam locomotive. The L&YR interest in Flamme's work centred on his novel 4-6-2 passenger and 2-10-0 freight locomotive designs that he had produced in 1910.

The year 1913 commenced with a shock for the whole family. Dr Wright at Horwich had concerns that Frances might have contracted consumption and had referred her to Dr Latham, a London specialist. He ordered her to be isolated for up to three months and she was admitted to a sanatorium at Nordrach on Dee, Banchory, near Dundee, travelling there by train via Edinburgh on 17th January. The good news was that she was over her cough within a week and everything looked hopeful as she had been diagnosed so early that it would be curable. However, it took until 9th May until she was fully recovered. In the meantime, Bay Smythe, who had gone back with the family to *Overdale* on 27th December after their Christmas visit to Killiney, stayed on to take charge of domestic matters.

During his visits to see his wife, Eoghan would take the opportunity to fish the Dee at Banchory, and it was there that he actually caught his first salmon. On a visit in April the doctor advised that it would be madness to take Frances back to Horwich until after the following winter. He spoke with Aspinall about the matter and found that he had no objection to him living away from Horwich. The chief problem from Eoghan's point of view was that he would have to leave home by 7.45 a.m. and would not get back until 7.15 p.m. In a letter to his mother from *Overdale* he made some other very interesting comments:

> On the other hand, I shan't have a factory chimney blowing smoke into my bedroom window and suffocating me, as was the case the other night. I certainly shan't leave the L&YR unless I can get a better and as permanent a job elsewhere, and I really wouldn't care to move for another year at least, much as I dislike this climate and the situation of this house, for I delight in my work. I hope that some day I'll be CME of the London & South Western Railway or one of those lines.

Staying with personal matters, another family connection with the world of railway electric traction was established on 18th June 1913 at St. Michael's church, Lyndhurst, Hampshire, when Frances' first cousin, Stella Alice Pauline Byrne deSatur, was married to Charles Hesterman Merz, one of the founding partners of the renowned electrical engineering consultants, Merz & McLellan. By early

CHAPTER SEVEN Management of Workshops

August Eoghan had arranged to lease a property at 35, Westbourne Road, Birkdale, near Southport, and Murrough and Ellen were able to see it when travelling through England on their way to Koblenz, Rheinfels and Weisbaden via Hook of Holland. While there, Murrough took ill again and was admitted to hospital in Switzerland where a Prof. Schüle thought that his case was serious. When he arrived home he consulted Dr Gordon who arranged to operate on the 16th September – his 40th wedding anniversary – when an entrogastronomy was performed, and a fortnight later he was home at *Mount Eagle*.

Writing to his parents in October from his new home at Birkdale, Eoghan described his daily routine:

> I get up at 7.0, breakfast at 7.30, leave the house at 8.0, and arrive at the station at 8.10 a.m. I buy the *Daily Citizen*, *Daily Mail* and *Liverpool Daily Post* and travel in the Guard's Van to Wigan. Then I get on the footplate, give the driver the papers, and have a chat with him as far as Lostock where I change. I wait there for ten minutes and chat with the stationmaster or my friend Mason, who is a manufacturer of firebricks and a sturdy Liberal. The 9.0 o'clock Horwich train comes in drawn by a locomotive recently repaired at the works. I travel on the footplate and chat with the driver and reach my office by 9.25. Office works 9.30 to 1.0 p.m., then lunch in the waiting room if Hughes is not there, which is often. Occasionally I go into Bolton. From 2.0 to 5.0 I am usually in the Works and from 5.0 to 5.30 I sign letters. I catch the 5.40 train to Lostock, change there and wait five minutes. I read reports from my assistants which I have not had time to read in my office, and arrive at Southport at 6.45 p.m. I cycle back up to the house, have dinner at 7.0 and to bed at 9.0 or 9.30, and so each day passes. Always I am kept very busy. I am just sending three of our foremen and draughtsmen to Belgium, and I correspond a good deal with Mr Flamme, the administrator of the Locomotive Department of the Belgian State Railways.

Eoghan O'Brien photographed with his mother Ellen at Mount Eagle in 1913. - [M.J. O'Brien; courtesy Prof. O.F. Robinson]

Frances, Brian and Eoghan at 35 Westbourne Road, Birkdale, in 1914. - [Courtesy Prof. O.F. Robinson]

With the construction of the superheated 0-8-0 goods locomotives proceeding apace, Hughes had written to Aspinall on the subject of single heading versus double heading of goods trains. In a letter to Hughes on 25th September 1913 H.E. O'Brien added his views on the matter. He felt that the rapid development of road transport made it all the more important that the L&YR should give quick delivery of any goods that they dealt with. Consequently, he felt that there was a need for bigger and faster goods locomotives, but stressed that if such big goods locomotives were built they should be worked with fully loaded trains. He pointed out that the double-headed trains then in operation were not fully loaded as a rule and therefore the charges for wages, interest and repairs were being spread over a smaller number of tons hauled than should be the case. He concluded by adding that not only would a larger locomotive be able to haul the heavy trains, but it would also be able to work the smaller vacuum braked goods trains at much higher speeds.

Murrough O'Brien took ill again early in 1914 and he left *Mount Eagle* to go into *Mount Elpis* private hospital for an operation on 12th February, but he was never to return. The surgeon considered that the days following the operation would be very anxious ones. Eoghan travelled to Dublin immediately to be with his father, but on Saturday 21st the doctors felt that another operation was the only thing that might save him. Regrettably, it was not successful; he was dying, and both Eoghan and Ellen were with Murrough when he passed away on the next day. Eoghan was therefore not in a position to attend a paper on *Some railway conditions governing Electrification* read at the I.E.E. on 12th February 1914 by Roger T. Smith, Electrical Engineer of the GWR. However, he did submit a lengthy written contribution to the discussion. In it he expressed the view that the operation of fast passenger services by electric locomotives was still only a vision in the United Kingdom. He supported his view by stating that, improvements in boiler design, the adoption of high-degree superheat, and the introduction of four-cylinder balanced engines had placed the steam locomotive far ahead of the electric one for some time to come. He added that there were specialised services on nearly every railway that could well be profitably worked by electric traction, but that each case had to be carefully considered on its merits. In responding, the author expressed his gratitude in finding that H.E. O'Brien was on the whole in agreement with the principles laid down in the paper.

Neither was Eoghan able to attend at the I.E.E. in March when Francis Lydall read his paper on *Motor and Control Equipments for Electric Locomotives*. Once again he contributed to the discussion through a written communication, noting that it was very desirable that the control of any electric locomotive should be entirely by contactors, and that the contactors should be placed in such a manner that noise generated by any activity taking place in them could not affect the driver. He added that the geared-motor bogie was the most suitable type for railway application and the one most likely to offer low maintenance costs.

However, he was at the I.E.E. on 12th March 1914 to read his own paper on *Design of Rolling Stock for Electric Railways*. He set the scene for his topic by noting that railways engaging in electrification of large suburban networks found that the cost of operating the electric service was an important proportion of their total expenditure. It was therefore a matter of importance to use every means to reduce the cost of operation. He pointed out that the cost of electrification and its subsequent operation depended on three factors: (1) the number of stops per mile; (2) the scheduled speed of the trains; and (3) the weight of the trains, noting that factor (1) was fixed by the nature of the district served and that factor (2) was a matter for management, adding that the main object of electrification was, of course, increased traffic. H.E. O'Brien was also at pains to draw attention to the fact that a profitable electrification often involved runs of very short distances between stops in the vicinity of large towns, and runs of much longer distances as the trains got farther from the urban areas. Other operational features that he referred to were variations in gradients and the requirement to run both express and stopping services on the same route. The point of his paper was that rolling stock had to be designed to meet these situations.

Drawing on his experience on the L&YR, he went on to detail the various design features that had to be addressed, including weight, fire resistance, seating capacity, vehicle length, ventilation, entrance doors, roof structure and materials for construction. Under each heading he discussed the weight and cost implications. Turning to the running gear, he referred to two types of bogie; those with equalising beams between the axles and those built on the lines of the ordinary English carriage bogie. He expressed the opinion that the equalising beam type of bogie, always being the heavier type, was not suitable for on use on English railways with their excellent permanent way, but he felt that it might be desirable on American railways with poor track beds. Particulars of motor and trailer cars used by the leading electric railways in England and the USA were presented in a series of tables which accompanied the paper, and drawings of carriage ventilations arrangements, a 10ft wheelbase motor bogie, together with underframe weight distribution diagram, were also included.

During the ensuing discussion on the paper H.E. O'Brien, responding to points raised by W. Casson and various other contributors in relation to the wisdom of basing cost of repairs on vehicle weight, pointed out that his experience in the North of England tended to show that the repairs did depend very largely on the tonnage of the cars. In reply to a further point raised by W. Casson regarding the small seats provided in the L&YR stock, he suggested that Mr Casson should pay a visit to the North, adding "it is a rather out-of-the-way place and difficult to get to, but I think that if he will get into these cars he will find they are not so uncomfortable as they may seem".

Responding to Harold William Firth (Electrical Engineer, Great Eastern Railway) and R.H. Burnett's questions on the long-wheelbase bogies, he informed them that they were designed partly to minimise the rail wear and partly to accommodate larger traction motors. He added that these objectives were not fully achieved by the design as the increased weight of the traction motors negated the benefits of the longer wheelbase, and that wheel wear (and presumably rail wear) was just as great as on the shorter wheelbase bogies previously used. He considered that in order to design a satisfactory power bogie ample journal bearing surfaces and ample bearing areas on the faces of the axle-boxes were required so that there would be no heavy wear on the horns. In his view, the design of power bogies was a question of selecting the right material and putting it in the right place!

H.E. O'Brien was again in discussion with H.W. Firth on 23rd April when he attended his paper on *Electrification of Railways as affected by traffic considerations* at the I.E.E. Noting that the paper dealt specifically with London conditions, he felt that it was only with considerable diffidence that those involved with provincial railways would venture to enter into discussion on it. However, he ventured to comment in regard to the main question that had been addressed, the working of peak load traffic. He was of the view that peaks were inherent in the nature of railway urban business and that there were two aspects to the matter – rolling stock and track, and power station plant. Although the author had apparently suggested the use of steam locomotives at peak times to smooth out the demand on the power station and for electric plant and rolling stock, H.E. O'Brien felt that the appropriate way to address the problem for the power plant was by the use of batteries, which not only eliminated momentary and hourly peaks but were also a valuable insurance and standby.

Agreeing with the author that the provision of rolling stock to meet peak demands was a more difficult matter, he noted that it had not been stated in the paper what might be done with the steam locomotives outside of rush-hour periods. H.E. O'Brien was of the opinion that it was preferable to have motor cars of sufficient power to haul the maximum number of trailers required at peak traffic times, which could be used with fewer trailers during lower traffic periods. He did concede that there might be some specialised cases for steam-hauled suburban services in London and elsewhere, and in that context he referred to the superheated four-cylinder 2-8-2 tank locomotives that the Belgian State Railways had designed specifically for suburban work. However, he did note that with the locomotive always coupled to one end of the train BoT approval would be required for similar push-pull operations.

In replying, H.W. Firth considered that, in relation to improving the load on the power plant, a great deal could be done by the traffic officers but that the kind of service that had to be operated depended on the nature of the district served. On the matter of varying the number of trailer vehicles, between peak and off-peak times, he felt that it might not entirely meet the case in respect of intense peaks. With regard to the use of superheated steam for suburban locomotives he expressed

CHAPTER SEVEN Management of Workshops

interest in the remarks "from a representative of a railway that had done so much in regard to superheating", but thought that its application to suburban services was a difficult subject.

In December 1913 the L&YR Board had authorised the electrification of the 9½-mile Manchester-Bury line. This was to be the first stage in a comprehensive electrification scheme for suburban lines in the Manchester area. As it was felt that installation of overhead line equipment on the complex trackwork layouts at the city end would prove unduly difficult, a shielded third rail system operating at 1,200V DC was adopted. A new Power House was constructed at Clifton Junction, but as this was not on the route it was necessary to run the feeder cables along the East Lancashire line between the Power House and Radcliffe Bridge station. At the Power House there were two 5,000kW turbo-alternator sets, which ran at 1,500 rev/min and gave a 25Hz output at 6,600V. Superheated steam at 700°F, 200lbf/in^2 was supplied from Babcock & Wilcox boilers equipped with chain grate stokers and Green's economisers. There was also a Westinghouse 500kW auxiliary turbo-alternator to provide internal power supply for machinery and lighting. The construction work on the whole scheme was slowed down due to the War, and Eoghan was actually away on Government service with the Ministry of Munitions when the Manchester-Bury line was eventually brought into operation on 7th April 1916. With the opening pending, A. Lund had been appointed Resident Electrical Engineer, Clifton Junction, from 24th July 1915.

On the electrified network in the Liverpool area, the live rail had been installed on the short single-track section from North Mersey Junction to Gladstone Dock, and services commenced on that line from 7th September 1914. However, they were to be relatively short-lived, ceasing on 7th July 1924. A spin off from the Manchester-Bury electrification was conversion of the Bury–Holcombe Brook branch to 1,200-V DC operation. Dick, Kerr & Co. had completed their experiments, and a major breakdown in the Holcombe Brook sub-station had required the introduction of a novel form of working. To enable the service to be maintained, power at 1,200V DC was fed into the overhead line from the Bury end of the branch, and one of the experimental motor cars was used to pick up current through its pantograph, otherwise operating simply as a trailer. Jumper cables were run to a Manchester motor car, which was coupled to it, so that appropriate traction could be provided for the lower voltage.

In sharp contrast to the Liverpool area electric stock, where so many varieties of car were operated, only three types of vehicle were constructed for the Manchester-Bury service; thirty-eight third-class motor cars (Diagram 136), fourteen third-class control trailers (Diagram 137), and fourteen first-class control trailers (Diagram 138). They were all built at Newton Heath to the same basic design, each car being 63ft 7in. long. Learning from the experience of the Marsh Lane accident, wood was eliminated from the design, an all-metal construction of aluminium panels on steel framework being employed to render the cars virtually fireproof. The motor bogies, which had outside springing, were constructed at Horwich. Twenty-eight motor cars were completed in 1915-16 along with nine third-class control trailers and nine first-class control trailers. An additional twenty vehicles were constructed to the same Diagrams during 1920-21

Following installation of the 1,200V DC third rail on the Holcombe Brook branch, services resumed on 29th March 1918 using Manchester-Bury rolling stock. The four experimental high-voltage vehicles were stored out of service until 1927, when the LMS authorised their conversion to a four-car diesel-electric set. New traction equipment comprising a 500hp 8-cylinder Beardmore diesel engine[59] coupled to a 340kW, 600V DC English Electric generator was installed in one of the former motor cars. After a period of trials the set was put into service between Blackpool Central and Lytham on 25th July 1928 and later it operated between Blackpool Central and Kirkham. However, the diesel-electric unit failed to perform with the anticipated reliability and it was withdrawn from service on 18th April 1929.

A five-car set of the 1916 Manchester – Bury electric stock - [HOR-F-1666 © NRM, York]

---

59 This was same model of diesel engine as used in the ill-fated airship R101

With the Royal Engineers

In the early months of the War the backbone of the system for supplying the British Expeditionary Force (BEF) was the existing network of standard-gauge railways in north-east France. The routes available were determined by the position of the front line established when the Allied forces managed to halt the German advance. The lines of communication were the two southbound railway routes that stemmed from the main supply ports of Calais and Dunkerque. The route lying further to the west was the excellently equipped Calais to Paris double-track main line, but the eastern route really consisted of a combination of main line and single-track cross-country branches. A main line ran from Dunkerque to Hazebrouck, which was the centre for the northern sector of the British front line. At Hazebrouck a branch from Boulogne came in from the west and branches to Armentières and Ypres ran off to the east and north-east respectively. The main line followed a circuitous route to Béthune, south of Hazebrouck, where it then turned in a south-westerly direction toward St Pol, Frevent, and Abbeville. At Frevent another line branched off to the south to connect with the important railway town of Amiens.

At first all lines were operated by the French, with priorities for military and civilian traffic decided by the military authority established by the Allies[60]. Once the front line had been established, it became imperative to provide additional facilities to cater for the tremendous increase in traffic. It was soon realised that the Royal Engineers (RE) had insufficient capacity in its two regular Railway Companies[61] to meet the increased demand for railway construction work, and the Director of Railway Transport was therefore instructed to organise additional Railway Construction units. The Railway Executive Committee had established a sub-committee for recruiting in October 1914 and men for the new RE Companies were selected from the large numbers of railway employees who were volunteering for military service. Approximately half of the officers required came from British railway companies on the recommendation of the Railway Executive Committee and the other half were mainly men from overseas who had been employed on colonial and foreign railways.

Included among their number was Eoghan O'Brien who, on 8th October 1914, took a temporary commission with the Royal Engineers with the rank of Captain. In his absence from Horwich George Nuttal Shawcross was appointed to act as Works Manager from 17th November. Ellen went over to stay with Frances and Brian at Birkdale, but on 16th November they left their Lancashire home. After spending three days at the *Artillery Mansions* hotel in Westminster, they took up residence at the *Royal Anchor Hotel* in Liphook, Hampshire, so as to be near Eoghan who was

---

60 Commission de Reseau pour les Chemins de Fer du Nord
61 8th & 10th Railway Companies RE

based at Longmoor Camp. Longmoor was the HQ of the regular railway troops before the War and, together with a part of the nearby Bordon Camp, this became the centre for all railway and road transport personnel for the duration of the War.

Capt. O'Brien with E.L. Vaughan (his housemaster at Eton) before his departure for France in February 1915 with 111th Company RE - [Courtesy Prof. O.F. Robinson]

A poor image, but of historic significance: 111th Company RE departing from Longmoor for northern France in February 1915 - [Courtesy Prof. O.F. Robinson]

It was at Longmoor Camp that Capt. O'Brien underwent a short spell of military training in preparation for taking up his first command as Officer Commanding 111th Company Royal Engineers, which was raised from amongst the railwaymen who had also been undergoing training at Longmoor since their recruitment. On

Officers and NCOs of 111th Railway Coy RE: Lieut A.H.M. Campion; Capt. H.E. O'Brien, Lieut G.W. Phillips, Lieut J.V. Nimmo, and Lieut S.E. Fay are seated centre front (4th to 8th from left - [Courtesy Prof. O.F. Robinson]

15th February 1915 they departed for the France & Flanders field of operations, and the journey by train to their port of embarkation was excellent. Having arranged refreshments for their men, the officers went into the town for lunch where, for what was his last meal in England for some time, Capt. O'Brien had oysters, grilled salmon and rhubarb tart and some excellent coffee. They had a fine crossing to

CHAPTER EIGHT **With the Royal Engineers**

Officers of the Railway Companies of the Royal Engineers stationed at Longmoor

Back row – Lieut J.P.S. Greig, Lieut V.O Winter, Lieut G.G. Glyn, Lieut G.W. Phillips, 2nd Lieut P.D. McFeat, Lieut E.A. Wilson, Lieut A.H.M. Campion, Lieut S.E. Fay, Lieut L.C. Reid, Lieut J.V. Nimmo; second row – Lieut A.E. Hallinan, Lieut C.H.W. Edmonds, Lieut P.R. Jordi, Lieut E.V.C. Wellesley, Lieut C.R. McIver, Lieut C.J. Newbold, Lieut S.H. Foot, Lieut R.O. Squarey; Lieut J. Briggs; front – Lieut C.E. Jordan-Bell, Lieut J.P. Rhodes, Capt N. Wilson, Capt H.P.M. Beames, Capt W.G. Tyrrell, Col H.M. Sinclair (CO), Capt. R.H. Greig DSO (adjutant), Capt F. Forbes-Higginson, Capt P.D. Michod , Capt H.E. O'Brien, 2nd Lieut R. McCreary

[Courtesy Prof. O.F. Robinson]

France, and the War Diary of the 111th Company RE records that they arrived at Abbeville on the morning of 18th February, where the officers and other ranks were billeted at the offices of Messieurs Saint Frères, Rue du Havre.[62] In getting his men well looked after Capt. O'Brien's fluency in French would have been of enormous help. His officers included Capt. Watts and Lieut. G.W. Phillips. Three other officers, Lieutenants A.H.M. Campion, S.E. Fay and J.V. Nimmo, had been temporarily relieved from Company duties and were working to instructions of the Railway Construction Engineer (RCE).[63]

Capt. O'Brien's first duty was to prepare a Construction Train for the accommodation of his Company and for their transportation with all necessary tools and equipment to and from wherever work was required. With this in mind he set off on the morning of 21st February to examine the design of the 8th Company's train, which was about twenty miles away. Having seen the train he went a further three miles to meet with Col. (later Brig-Gen. Sir William) Waghorn and Gen. J.H. Twiss, before returning to the 8th Company's train where he dined with four of their officers and stayed the night. The following day

Capt. H.E. O'Brien, in command of 111th Company, Railway Troops, Royal Engineers

---

62 PRO WO95/4053

63 George William Phillips (1883-1971) was trained as a mechanical engineer in the GNR workshops at Doncaster and appointed District Locomotive Superintendent, York, in October 1912, MIMechE and MILocoE. Attained the rank of Major on staff of Assistant Director of Light Railways; and was L&NER District Locomotive Superintendent, Glasgow on his retirement in December 1943.

Arthur Havard Montriou Campion (b1887), a graduate of the University of London, was appointed an assistant engineer on the Bombay Baroda & Central India Railway in March 1909, AMICE. He attained the rank of Major with the staff of the Assistant Director of Railway Construction; and by 1942 was Deputy Engineer-in-Chief for the BB&CIR metre-gauge system.

Samuel Ernest Fay (1891-1987), son of Sir Sam Fay, General Manager of the Great Central Railway, gained varied practical experience of railway operations in England, Europe, Canada and America, MInstT. During the war he acted for a time as Deputy Assistant Director-General of Movements and Railways. He was subsequently appointed GCR Assistant District Superintendent, London Division, and was Chief Operating Officer of the São Paulo Railway, Brazil, before taking an appointment as Operation and Equipment Assistant for the New Zealand Railways.

James Valence Nimmo (1875-1948) served an engineering apprenticeship with John Fowler & Co., Leeds, gained practical experience on the Assam Bengal Railway (1897-1901) and graduated from Glasgow University in 1902. He was next engaged on railway contracts with S. Pearson & Son and worked in Canada as Resident Engineer for the Atlantic Quebec & Western Railway (1904-07). Appointed Engineer for the Northern Pacific division of the Canadian Northern Railway in 1909. MICE 1912, and also member of American Society of Civil Engineers, Canadian Society of Civil Engineers, American Railway Engineering Association, and American Society for Testing Materials

Map showing operational area of 111th Company RE in northern France

Capt. O'Brien went to see the 109th Company's train, where he met Lieut. Jordan-Bell and lunched with him. During the same week, the whole of the Company, with the exception of one section employed on Barrack duties, were put to work on the sidings at Abbeville station which were part of the facilities for the First Army's advance base depot and regulating station. It was here that trains arriving from the Channel ports were re-marshalled into formations containing the quantities of supplies required by each Division and then forwarded to the railheads that had been established within reasonable distance of the front line.

During the next week of 1st to 6th March Capt. O'Brien was busy preparing designs and arrangements for fitting-up the Construction Train, the work of which was put in hand immediately and the twenty-five wagons were completed during the following week. On 7th March all of the officers, excepting Capt. Watts, left their billets at Messieurs Saint Frères factory and took up residence in the Construction Train which was placed on a siding at Abbeville station. Capt. Watts and the Company joined them in the train on 8th March and two days later each man had been provided with his own bunk. On 9th March the YMCA erected two tents down by the sidings, one for concerts and the other for reading and writing. The first concert was held on the evening of 8th March, which Capt. O'Brien attended along with Capt. H.P.M. Beames.

Meanwhile, work on the sidings at Abbeville station continued; on Saturday 13th and Sunday 14th March Capt. O'Brien visited quarries at Noeux, Bully-Grenay and Marles with a view to accelerating the delivery of ballast for the new sidings at Abbeville. Work was advancing rapidly, but as the sidings were being laid on boggy ground a large supply of ballast was needed. The following week he went to watch a steam pile-driver in action driving piles for the foundations of an electric power station. As it was just what was required for bridging work, he spent two days receiving instructions in its use and on the method of manufacturing concrete piles. Meanwhile, in addition to supervising the work at the sidings, Lieut. Phillips was looking after the motor lorries and motor bicycles that had been delivered to the Company.

On the afternoon of 24th March some wagons became badly derailed, the Brake Van sinking into the boggy ground up to its floorboards. A civil engineer was called in to re-rail the vehicles while Beames, Phillips and O'Brien observed the proceedings with an appropriate degree of cynicism. The civil engineers were still at the job the next day and they took twenty-six hours to re-rail the Brake Van. As a result of that fiasco, Capt. O'Brien was made O.C. for all derailments with Lieut. Phillips as his assistant, and he was to record, rather wryly, that this meant that they would have to put on absolutely anything that came off the rails. In a more pleasant moment he went to see a large French locomotive in the afternoon of 27th March and talked with the fireman.

Capt. O'Brien went to Blangy on Sunday 28th March with one of the motor lorries to bring back a 3-ton road roller, which he had arranged to borrow from Ingenieur du Ponts et Chausées.[64] The next day was spent in attaching the roller to the motor lorry in order to make it ready for use in rolling a roadway to the Supplies Depot at the new railway yard at Abbeville, the road rolling work being thereafter supervised by Lieut. Phillips. In accordance with the Officer Commanding Royal Engineer's instructions, two Army Service Corps' men attached to the Company, Pte J. Ross and Pte J. Flint, were transferred to 109th Company, RE, on 8th April together with one 3-ton motor lorry, one motor bicycle and one bicycle. On Friday 9th Major F.A. Twiss visited the Construction Train and enquired into the method of dealing with Pay Books, returns, etc. A draft of reinforcements (2 NCOs and 17 Sappers) arrived at Abbeville on 11th April, thus bringing 111th Company up to full strength.

Another derailment took place late on 13th April when one of the men from the Royal Monmouth Regiment, who was supervising the shunting, allowed a locomotive to run off the end of one of the sidings and into the soft ground. Capt. O'Brien was called out of bed but, when he saw that it was up to its buffer beam in the mud and not in the way of anything, he decided to leave the re-railing till the morning light. He was up at 5.30 a.m. and with Capt. Philips organised a recovery gang which worked for twelve hours getting the locomotive back on the road. They celebrated their success that evening with a magnificent cake, a present from Mrs Aspinall that had arrived the same day. Two days later another concert was held in the YMCA tent at which the two Miss Bowen-Cookes, daughters of the CME of the L&NWR, were in attendance. They had only just arrived at Abbeville, and as well as singing in the concert, they also helped to prepare and serve teas.

With work on the new sidings at Abbeville almost complete, and the decision having been taken that the British would double the line from Hazebrouck to Ypres, RCE2 sent a telegram to 111th Company instructing them to make arrangements to move from Abbeville to Arques (Balastière). The Company worked as usual until 1.0 p.m. on 25th April, and on parade at 2.0 p.m. they were addressed by Major N. Wilson, RE, Chief Railway Construction Engineer 2, who complimented all concerned very highly on the work they had performed at Abbeville and gave

---
64 Bridges & Causeways Engineer

them a holiday for the remainder of the day. At midnight 5 Officers and 247 other ranks left Abbeville in the Construction Train bound for Arques, arriving at their destination at about 8.0 a.m. the following morning. One officer together with four other ranks left Abbeville early on Monday morning by motor lorry. The Company paraded in Arques at 9.0 a.m. and were then dismissed for the day, excepting those required for Company duties. Authority was granted on 25th April for Sapper J. Nicholson, who had been acting L/Corporal since the Company had embarked for France, to be promoted to that rank with pay back-dated to 15th February 1915.

Owing to the transfer of the 111th Company to RCE1 section, certain re-arrangements of the Construction Train were necessary. In collaboration with the Acting Adjutant and Officer Commanding 109th Company RE, Capt. O'Brien ensured that these were carried out during the afternoon of 26th April. On the following day, No. 3 Platoon, comprising Lieut. Nimmo and 59 other ranks, was instructed to prepare to leave for Wizernes to join 8th Company RE in No.2 Train, which they did at 11.0 a.m. Orders were also given for 50 NCOs and men of No.1 Platoon to prepare to leave for Caestre on Wednesday 28th, which they did in two detachments by motor lorry at 10.0 a.m. and 2.30 p.m. The remaining two platoons were employed on the following two days on general straightening up of the sidings at Arques and in re-fitting the trucks due to the re-arrangement of the Construction Train.

The marshalling yard on the Hazerbrouck - Armentiers line. - [© IWM Q46767]

On 30th April arrangements were made for the No.1 Construction Train to leave for Caestre the next day. One officer and fifty other ranks of No.2 Platoon left Arques for Caestre by motor lorry on 1st May whilst No.1 Platoon finished straightening up the sidings, which work was completed at 4.0 p.m. At 6.45 p.m. they also left Arques for Caestre, arriving at their destination at about 11.0 p.m. and together with No.1 Platoon rejoined No.1 Construction Train, which had been moved separately to Caestre on the same day. Between 2nd and 22nd May the weather was fine and warm, and the officers and three platoons of the Company were engaged on widening the Caestre to Godewaersvelde section of the Hazerbrouck to Ypres line, according to instructions of Railway Construction Engineer 1.

Capt. O'Brien had met again with Col. Waghorn at GHQ on the evening of 27th April and during a long talk with him expressed his desire to be somewhere where he could do some good and useful work for his country. Little did he know when he left Hazerbrouck on the evening of 19th May 1915, having been granted leave to proceed to England, that his wish had been considered and that he was already earmarked for another and more important role with the Ministry of Munitions. A week later he was back at Horwich on temporary loan to the L&YR to assist with gearing up the railway workshops for munitions work, and on 28th June the 111th Company RE received notification that Capt. O'Brien would not be available to return for service with them.

Capt. H.E. O'Brien at Southport in May 1915 following his return from service in France with the Royal Engineers
- [Courtesy Prof. O.F. Robinson]

National Projectile Factories

By early October 1914 it was found that the Government arsenals and the half-dozen or so established private munitions works were quite inadequate to supply the vast amount of shells required by the Army — it had become obvious that an output of gun ammunition on an unprecedented scale was essential. During January 1915 a survey was undertaken of the nation's engineering resources and the concept of using local co-operative manufacturing groups was promoted, but Gen. Guthrie Smith, Director of Artillery, gave this chilly encouragement[65] — he was even unwilling to send any representatives to a demonstration in Leicester. The supply of munitions was a matter for the individual services, and as far as he was concerned that was the way it was going to stay for the Army. This narrow vision, by those responsible for supply of munitions, clearly gives the key to the whole policy that was pursued until April 1915.

The question of munitions supply, especially of high explosive shells, had become critical and it was obvious that the situation could not be tolerated any longer. So it was that on 8th April 1915 the Prime Minister, Herbert Asquith, announced in the House the appointment of a Cabinet committee on munitions under Lloyd George's chairmanship. During his speech Asquith made this very important statement: "The function of the Committee is to ensure the promptest and most efficient application of all available productive resources of the country to the manufacture and supply of munitions of war to the Army and Navy. It has full power to take all steps necessary for that purpose." The Munitions of War Committee worked through a War Office committee, the Armaments Output Committee, which had been set up a week earlier.

It was perfectly evident to the Munitions of War Committee that a separate Ministry, with the drive and freedom necessary to secure the mobilisation of the nation's entire engineering resources, was essential. The Ministry of Munitions, which was formed on the 9th June 1915, superseded the two committees established in April and assumed responsibility for the manufacture and supply of light and heavy arms, ammunition and explosives. Administration of the Royal Ordnance factories at Woolwich, Enfield and Waltham Abbey and any new munitions factories was to be under its control, but the naval dockyards were not. The Ministry's responsibilities also covered the matter of contracts and labour in the munitions industry.

Lloyd George was appointed Minister for Munitions with Christopher Addison as Parliamentary Secretary. In order to overcome the previous inadequacies and produce munitions quickly, it was vital to work as quickly as possible in setting up the new Department and their first duty was to recruit a team of experienced personnel to assist them in their great task. One of the first people to tender his

---

65 Addison Papers, C.23

services was Eric Campbell Geddes, Deputy General Manager of the NER. Lloyd George recalled the impression that Geddes made upon him on the morning that he rolled into his office: "He had the make of one of their powerful locomotives, and struck me immediately as a man of exceptional force and capacity."[66] Appointed to the position of Deputy Director General of Munitions Supply in June 1915, Eric Geddes had received his education at Edinburgh Academy, and Oxford Military College, following which he gained wide experience in railway operations with the Baltimore & Ohio Railroad and in India before joining the NER in 1904.

Working together, Addison and Geddes obtained the services of some other notable railwaymen. Assisting Geddes in a department, which was responsible for rifles, small arms, ammunition and machine guns, was Major Ralph Lewis Wedgwood, Chief Goods Manager and Passenger Manager of the NER. Henry Fowler, CME of the MR, was appointed to supervise the National Factories, and on 28th June 1915 Eoghan O'Brien, was appointed as Fowler's personal assistant at £500 p.a. with the rank of temporary Major. Geddes and Wedgwood had their offices on the 1st floor of Armaments Building, Whitehall Place, and Fowler and O'Brien were in Rooms 408 and 406 respectively on the 4th Floor —telephone Extensions 163 and 165.

This little coterie of railwaymen were well-known to each other; not only were Geddes and Wedgwood both from the NER, but one of their Directors, Murrough Wilson, was also O'Brien's first cousin. In addition, Fowler and O'Brien were long acquainted since they first met as teacher and pupil at the Horwich Mechanics Institute way back in 1898. Lloyd George was to record that "no more remarkable collection of men was ever gathered together under the same roof, adding that all the means of production, distribution, and exchange were aggregately at their command."[67] Two months later another railway engineer of note was to join the Ministry to head the organisation proposed to deal with the Royal Ordnance factories. On 23rd August 1915 Addison wrote to Lloyd George stating that Geddes was very anxious to have for his assistant in this matter Vincent Litchfield Raven, the CME of the NER. Addison noted that Raven had "a reputation of being able to get along with people and to get the best out of a factory", confirming that he had met him and liked the look of him, and adding that "he seemed a good type of man".[68]

The National Factories, for which Fowler and O'Brien had responsibility, were of four principal classes: National Shell Factories (NSF), that were managed by Boards of Management; two classes of National Filling Factories (NFF) respectively managed by either Boards of Management or by firms acting as agents for the Government; and National Projectile Factories (NPF), which were managed by

---

66 War Memoirs of David Lloyd George
67 ibid
68 ibid

major engineering firms on behalf of the Government.[69] As Director of National Projectile Factories Major O'Brien's duties officially included staff organisation and general supervision of all questions relating to electric generating stations and power supply, but he was soon involved with all aspects of the department. He was also Chairman of the NPF Committee and was at the conference held on 18th June 1915 at 6, Whitehall Gardens, when it was decided that the U.K. should be divided into ten administrative areas, viz.: England & Wales, 7 Areas (Newcastle, Leeds, Manchester, Birmingham, London, Bristol and Cardiff); Scotland, 2 Areas (Edinburgh and Glasgow); and Ireland, 1 Area (Dublin).[70]

The Area Offices exercised supervision over the Local Boards of Management within their areas, and were each provided with an Organising Secretary, a Labour Officer, and a Superintendent Engineer. The Superintendent Engineers were charged with developing the resources of their Areas as fully as possible along the lines laid down, from time to time, by the Ministry of Munitions; ascertaining details of and reporting on available machinery; inspecting National Shell Factories, advising on capabilities of firms; and reporting on the progress of contracts. Their reports were sent to Major O'Brien, and Lloyd George chaired meetings with them at which both Fowler and O'Brien were in attendance. The Superintendent Engineer for the Leeds area was Oliver Winder, one of H.E. O'Brien's former colleagues from his days at Horwich.

On 27th June 1915 Major O'Brien and his family left Lancashire and drove to London, stopping overnight at the *Randolph* hotel in Oxford. On arrival in London, they were at first provided with accommodation at *The Lodge*, St. James' Park, Westminster, for one month, but then a more permanent home was arranged at 70, Warwick Square. Apart from the times when the family visited Ireland, this was to be their principal residence until the end of November, but from 15th December until the 14th March 1916 they lived at Pyrford, near Woking in Surrey.

At a meeting held on 24th July 1915 Addison proposed that in order to spread the base of supply of what he termed 'the rougher munitions of war', such as shells, and to make the most use of Government capital, the services of the greater proportion of high-calibre engineering firms should be requisitioned. In that way their skilled management and staff could be employed to the utmost to produce the more easily manufactured munitions. Amongst such firms would be, he submitted, the large railway rolling stock manufacturing workshops throughout the country such as: Andrew Barclay (Kilmarnock); Beyer Peacock & Co. (Gorton); the Birmingham Railway Carriage & Wagon Co.; British Westinghouse Electric (Manchester); Brush Electrical Engineering (Loughborough); Davey Paxman Ltd. (Colchester); Hudswell-Clarke (Leeds); Hunslet Engine Co. (Leeds); Hurst Nelson (Motherwell);

---

69 Addison Papers, C37
70 ibid

the Leeds Forge Co.; Manning Wardle (Leeds); the Midland Railway Carriage & Wagon Co. (Birmingham); North British Locomotive Co. (Glasgow); Patent Shaft & Axletree Co. (Wednesbury); Robert Stephenson (Newcastle-upon-Tyne); and J. Stone (Deptford).

Capt. O'Brien and son Brian at Mount Eagle in 1915 - [Courtesy Prof. O.F. Robinson]

70, Warwick Square, London, where Major H.E. O'Brien lived with his family from August to December 1915 while working at the Ministry of Munitions - [Courtesy Prof. O.F. Robinson]

Major O'Brien also served on the Departmental Area Organisation (DAO) Executive Committee under the chairmanship of a Mr Stevenson.[71] At its 20th Meeting, held on 29th September 1915, Major O'Brien reported that he had the operations of the NSF at Liverpool Haymarket under close observation by an inspector and that everything possible was being done to reduce the costs. At the same meeting he reported that the Birmingham Board of Management had come to him with a scheme for extending the generating and other plant at the Birmingham NSF, and that he had referred the matter to the Committee. The Secretary was instructed to write to the Birmingham Board and ask them to supply full particulars and at the next meeting, on 6th October, the application from the Birmingham Board for an expenditure of approximately £8,000 on additional power and hydraulic plant at the Birmingham NSF was tabled. The Committee decided to refuse sanction for expenditure until the lease of the workshops comprising the NSF had been agreed between the Birmingham Board and the owners of the premises, the Midland Carriage & Wagon Co. It was agreed that Mr McLaren and a representative of Major O'Brien's department should visit Birmingham to go into the whole question of this lease with the Board and arrive at an arrangement in regard to it that they could recommend to the Finance Department for acceptance.

Major O'Brien also reported at the meeting held on 6th October 1915 that he

---
71 Addison Papers, C37

CHAPTER NINE **National Projectile Factories**

endorsed the recommendation of the Leeds Area Superintendent Engineer, Oliver Winder, that a new manager should be appointed for the Bacup & Rawtenstall NSF. It was decided that the Area Board and Superintendent Engineer should be asked to come to London to discuss the matter with a view to change in the management. The question of the percentage of women and boys employed at NSFs, as recorded in a return dated 23rd September, was also discussed. The Committee decided to appoint a sub-committee, comprising Major O'Brien, Mr Maclean, Mr Lee Murray and Mr McLaren to arrange interviews with those Boards of Management of NSFs who were employing too large a percentage of men, with a view to getting it reduced.

Important as all these administrative aspects were, it was in connection with one of the most remarkable industrial expansion projects — the National Projectile Factories — that Eoghan O'Brien was to make his most important contribution at the Ministry. Addison's initial idea to requisition the services of the large high-calibre engineering firms was not ideal for this purpose. Although such firms had the machine tools capable of producing anything asked of them, it was realised that the specialised work of heavy shell production could be carried out with greater rapidity and ease on machines specially designed to deal with shell manufacture and nothing else. Furthermore, the wide-ranging capabilities of the engineering firms' machine tools would be rendered ineffective if totally devoted to shell manufacturing processes. Labour considerations were also taken into account: to use the highly skilled machinists on straightforward production work would have been a waste of their skills, and there was already such a serious shortage that the Government had ceased to allow skilled workers to join up.

It was therefore decided to establish factories for the specific purpose of producing the heavier shell cases, and it was Major O'Brien's responsibility, as Director of National Projectile Factories, to oversee the project for the construction, equipment and start-up of twelve new factories. Well-known engineering firms, who had established for themselves a sound reputation for good workmanship, and who through the success of their own undertakings had shown that they knew how to organise and operate an engineering factory, were requested to select sites, erect buildings, order the machines, and take charge of the management. The firms concerned, and the location of the NPFs, were as follows[72]:—

Harper Bean & Son	Dudley
Dick Kerr & Co.	Hackney Marshes, East London
T. Firth & Sons Ltd.	Sheffield
Beardmore & Co.	Bridgeton Plant, Mile End, Glasgow
	Cardonald, Glasgow
Vickers Ltd.	Lancaster
Babcock & Wilcox	Renfrew
Leeds N.S &P.F. Board	Hunslet, Leeds

72 National Archives, MUN 5/146/1122/11

Cammell Laird & Co.	Nottingham
Hadfields Ltd.	Sheffield
Armstrong Whitworth & Co.	Birtley, Co. Durham
G.& J. Weir Ltd.	Albert Factory, Cathcart, Lanarkshire

Note: Beardmore built a separate Forge at Mossend to supply both Bridgeton and Cardonald factories.

In each case, the cost of purchasing or leasing the land, of erecting the necessary buildings and of supplying the machinery and other equipment was borne by the Government. They also paid fees for the design and supervision of erection of the buildings. The firms were responsible for ordering the machines required and having them installed and when all was working the Government provided the funds to meet the workforce wages and pay for management. This resulted in, what was for its time; the anomaly of a Government-owned arsenal being designed, staffed and run by an individual private firm — one of the earliest examples of a public-private partnership. In practice it all worked out extremely well and combined the best qualities of a Government factory with those of private enterprise.

By late August 1915 the sites had been selected and one of the first problems that had to be dealt with was the connection of the chosen site with the nearest railway for the delivery of construction materials. In some cases this involved very steeply graded connecting lines, which were also later used for the removal of the finished shells. Filling and levelling of the sites commenced immediately, two or three thousand tons of ashes being required at certain locations for this purpose alone. Work on pouring concrete foundations 3ft square by 3ft or 4ft deep, commenced in early September and by the middle of that month erection of the steel columns had begun, these being held in place on the foundations by bolts 18in. long. The roof trusses followed and by Christmas 1915 the steelwork was complete. A certain Mr Hunter was in charge of the steelwork erection at all the NPFs, which for a typical factory involved about 1,800 tons of steel in columns and roof trusses.

Bad weather at the beginning of 1916 delayed progress and it was not until around the end of February that glazing of the roofs was finished. The glazing of a typical roof required about 400 tons of glass, which was tinted blue by stippling on the inside with washable distemper to filter out the direct rays from the sun. Finally, the side-walls of the building were clad with galvanised iron sheets. The office walls were made from standard pre-cast concrete paving slabs, 2ft square and 2½ in. thick, which were ideal for the purpose. They were cemented together along their edges and each adjoining pair of slabs was bound together by small steel U-shaped pieces. Reinforced concrete pillars were used to stiffen the inside and outside of the walls at 29ft intervals and the outside of the walls was rough cast. The factory structures were completed around the end of March by which time the concrete foundations for the machines were ready.

One of the National Projectile Factories under construction, 27th January 1916. - [MUN5-290-001© TNA,Kew]

Meanwhile, day-to-day matters at the Ministry of Munitions still required attention; and even George Hughes turned to Major O'Brien for some guidance. In connection with his duties on the Horwich local tribunal Hughes had to visit the works of W.H. Pickup & Co. and John Crankshaw & Co., which were under the same management. At Crankshaw's they had miles of sanitary pipes on order for military camps, and at Pickup's they were very successfully making enamel ware for the conveyance of picric acid. In his letter of 20th March 1916, Hughes stated that if any more men had to go to the front from either works one of them would have to be closed. He sought a view on the relative importance of the two workshops so that if one of them had to close the correct choice could be made. Unfortunately, Major O'Brien's reply has not survived, and we do not know if enamel ware for the conveyance of picric acid was considered of greater or less importance than sanitary drain pipes for military camps.

Prior to this, at the end of November 1915, Major O'Brien had set in place specific procedures for the approval of orders and certifying of plant and machinery for the NPFs. Taking account of comments received from the firms that were engaged in the construction of the factories, he clarified matters in Ministry of Munitions Circular No.57, dated 17th December 1915. In this he emphasised that for Treasury and Audit purposes it was absolutely necessary that there should be an approval for every order given on behalf of a NPF, but in order to simplify matters as much as possible he suggested that the following procedure[73] should be adopted:

---
73 National Archives, MUN 4/476

1. That you should send one copy of each order to the Ministry (instead of two as asked for in my letter of 26th November) the second copy being sent direct to the Superintendent Engineer.

    (Note: In cases where a copy or copies of orders have already been sent to the Ministry, the Ministry will be responsible for sending a copy to the Superintendent Engineer.)

2. Approval of the order will be sent you after consideration by the Ministry. This approval together with the receipted account from the Contractor from whom you have ordered the goods or plant will be sufficient for the purpose of audit by Treasury.

3. That you will notify the delivery to you of any goods ordered by you on a form similar to the attached specimen, two copies being sent to the Ministry and one to the Superintendent Engineer, as far as possible confining the particulars on each form to a single order.

4. The Superintendent Engineer will from time to time examine the goods received and will report to the Ministry thereon, but it is not necessary for him to sign the notification of receipt to the Ministry.

5. This procedure relates to everything with the exception of buildings and machine tools, for which latter the procedure already adopted by the Machine Tool Department will be continued and letters must be addressed to that Department as heretofore.

    (Note: Machine Tools are to be understood to be generally all machines used for cutting, abrading or pressing metal.)

6. Approvals of all other plant and goods (buildings excepted) will be given by the Director of Production for the Director General, and all correspondence in connection therewith should be addressed to the Director of Production.

7. Any overlapping which may occur due to lack of a sharp line of demarcation between machine tools and other plant will be adjusted by the Ministry.

There was obviously still some misunderstanding amongst the engineering firms about what was meant by plant. Major O'Brien had to inform them in Circular No.67 that "it should be understood that this does not refer to equipment in the nature of small tools, general stores, or accessory equipment, the value of which is under £50." He went on to state that a separate account would have to be kept for these items, which would be audited from time to time, and he also took the opportunity to remind everyone that "The early completion of the factories is of the utmost importance and no work or orders must be held up pending receipt of approval".[74]

From the outset it was realised that the manufacture of shells could be carried out with greater efficiency on machines specifically designed to deal with only the specialised processes involved. It was also recognised that construction would be simpler the more specialised the machines were made and that it would be easier to train labour to handle and operate them. A certain amount of delay was inevitably incurred while the machines were designed and built, but in the end

---

[74] National Archives, MUN 4/476

the speed and capacity of shell manufacturing was greatly increased. The machine tool manufacturers set to work and, instead of making lathes capable of handling all 14 shell-machining operations; they produced 14 different types of machine, each one of which was specialised to do a single shell machining operation. The simplification of the machine mechanism in each case was carried out to a truly remarkable degree and as a result female and other workers with no previous acquaintance with an engineering factory, could be taught to operate the machines efficiently in anything from a fortnight to a month.

Naturally, the supply of the machine tools could not be accelerated beyond a certain limit, and Mary Humphry Ward[75] (who through the kindness of Lloyd George was enabled to visit the NPFs in the Midlands and on the Clyde in February 1915) was to record of some "I saw huge, empty workshops, waiting for their machines, or just setting them up, and everywhere the air was full of rumours of the new industrial forces – the armies of women that were to be brought to bear".[76] At other locations, where some of the machines had arrived before the end of 1915, section after section of the workshop had been set going some two months before the factory was complete and there she found "new workshops already filled with workers, a large proportion of them women, already turning out a mass of shell which would have seemed incredible to soldiers and civilians alike during the first months of the war".[77] However, things were progressing rapidly and at the beginning of April 1916 Major O'Brien was in a position to give the following detailed report on the NPFs:—

Harper Sons & Bean, Dudley

> The production of 60-Pdr shells is well in hand, but considerable efforts will have to be made, and the necessary labour supplied, if the output of 10,000 per week is to be given by the end of the month. Mr Bean is in close touch with the Labour Department with a view to obtaining this.
>
> The progress of the 6-inch plant is still delayed owing to want of millwrights, and efforts are being made to assist in this.

T. Firth & Sons, Sheffield

> This factory is more fully advanced than any other is and production is being pushed forward very energetically. The first 60-Pdr High Explosive (H.E.) shells will be delivered into bond this week. Good arrangements have been made for the supply of labour.

---

75 Mary Humphry Ward (1851-1920) was one of the most popular serious novelists writing in England in the late 1880s and 1890s. She was born Mary Augusta Arnold in Hobart, Van Diemen's Land. Her father, Tom Arnold, was the brother of Matthew Arnold. In 1872 she married Thomas Humphry Ward, whose name she took as a writer. Aldous Huxley was her nephew.

76 Humphry Ward, M., *England's Effort,* 1916

77 ibid

Beardmore & Co.

Forge, Mossend — This forge is progressing favourably and will give greater output than originally intended. It will be some little while before forging commences, but in the meantime the machining factories are being supplied by Messrs Beardmores from their own forge.

Bridgeton Plant — Progress here has not been quite as good as was hoped for. Arrangements for supplying certain necessary machines from other sources were made. The promise of delivery is 1,000 6-inch H.E. shells and 3,000 60-Pdr H.E. shells per week by the end of June, and should be improved on materially.

Cardonald — The supply of the larger size of forgings was discussed. The shop making 15-inch Howitzer shells was visited and the position discussed. In the 4.5-inch Shrapnel shop arrangements were made to expedite the change over to 60-Pdr Shrapnel.

Vickers Ltd., Lancaster

60-Pdr Shrapnel shop — progress is being made on this shell and an early output of this category should be obtained.

9.2-inch shop — there is a considerable amount of machinery here and as the determining factor of the output of this shop is the Power Station, and investigation proves that this will not be ready before end of May, it is suggested that certain of the 9.2-inch machines shall be transferred to shops where they can at once be utilised.

Babcock & Wilcox, Renfrew

60-Pdr Shrapnel Shop — one unit (1,000 per week) is starting this week and others should follow as quickly.

12-inch H.E. Machine shop and forge — progress here is behind all other factories, the shop itself not yet being finished.

Leeds NPF

Machining of 9.2-inch shells has commenced in this factory, but the shop is still short of lathes and chucks for certain operations. It is, therefore, proposed that these be obtained from Lancaster. The output with this assistance should be up to promise.

Cammell Laird & Co., Nottingham

Machining operations have now started on both 9.2-inch and 6-inch plants. With a little re-arrangement, it is felt that the final output of 9.2-inch can be achieved.

Hadfields Ltd., Sheffield

Operations on 9.2-inch shells have commenced here and it is anticipated that a considerable number of shells will pass the bonding stage this week.

G&J Weir, Ltd., Cathcart

6-inch Factory — progress here has been retarded by the lack of certain machines, which are already on rail from Lancaster NPF

8-inch Factory — There is a delay here owing to want of fitters and tool setters. This matter is being taken up.

One of the two NPFs not mentioned in the above report, that at Birtley, which was under the direction of Armstrong Whitworth & Co., may have been omitted as it was to be run with Belgian labour. While details of the NPFs were under consideration, three Belgian gentlemen approached the Ministry with a proposal to find sufficient Belgian labour, both skilled and unskilled, to run an entire shell factory. They agreed to carry out that proposal under the supervision of Armstrong Whitworth & Co., but in November 1915, when the construction of the factory was well advanced, some doubt was felt as to the ability of the Belgian gentlemen to procure the necessary labour. On 7th January 1916 the Ministry of Munitions held an interview with Baron de Broqueville, the Belgian Prime Minister, pointing out that it was highly desirable that the factory should be manned entirely by Belgians. As a result, the three Belgian gentlemen were dismissed and, at a conference that took place on 10th February 1916 in London, the terms of a Convention were fully agreed and signed by Lloyd George and Baron de Broqueville.[78] The headlines of the Convention were:—

1. Factory will be set up by British Government;
2. Manager will be nominated by Belgian Government;
3. British Government will supply all the machines, raw material and necessary capital;
4. Salaries and wages will be paid by British Government;
5. Foremen, skilled and unskilled labour will be paid according to rates ruling in that part of England;
6. Belgian Manager will commence work at beginning of February 1916;
7. Financial side of works will be undertaken by British Government;
8. British Government will nominate a representative who will control the English personnel and serve as intermediary between Belgian management and British Government;
9. British Government will try to arrange for Armstrong Whitworth to place at the disposal of the Belgian management its technical advice.

M. Hubert Debauche was subsequently appointed by the Belgian Government to be general manager of the factory, and a colony, known as Elizabethville, was specially established for the workers at the factory. Birtley NPF commenced production on 17th July 1916 under the control of Ministry of Munitions.

By 19th April, Major O'Brien was able to submit a table showing the comparative situation for the two weeks ending 15th April 1916 (see Appendix Seven). This gave details of the machines received, fixed and working out of the total of 9,697 that were to be installed. In addition, he showed that the number of employees had grown from 2,129 on 1st April to 3,689; 12.7% of the estimated 29,140 staff

---

[78] National Archives, MUN 5/78/327/24

Major H.E. O'Brien relaxing at Pyrford, where he resided from December 1915 to March 1916 while working at the Ministry of Munitions. - [Courtesy Prof. O.F. Robinson]

required for the 12 NPFs. Shells were starting to come off the production lines and, although only 691 had been completed, there were 32,657 in progress. The NPFs did not turn out shells complete and ready for handing over to the gunners; they simply took in steel billets, forged and machined them into shell cases, and made the steel nose cones that were screwed into the ends of the shells.[79] Neither were they involved with manufacture of the fuses that were screwed into the caps they made, nor with filling of shells with high explosives, which was a task that was performed at the separate National Filling Factories. On 22nd July 1916 the Chief of the Imperial Russian General Staff, General Belaieff, and his fellow officers visited the munitions workshop in the Wolsely works at Birmingham, and Major O'Brien was on hand for that most interesting experience which pre-dated the Russian revolution. The following details are from a description of the NPFs that appeared in *The Engineer*.[80]

> The NPFs were large buildings and generally comprised 42 rows of steel columns, which formed 41 working bays 18ft wide, each bay being provided with a pointed roof glazed on both slopes. The three centre bays were usually occupied by a blacksmiths' shop where the tools used in the machines were forged; a grinding shop where they were ground to shape and sharpened after being forged; a tool room where repairs to equipment for the machines was carried out; and a store room from which tools, oil and other items required by the operators were issued. Although the blacksmiths' shop was surrounded by the production area, no inconvenience was experienced, as electrically driven hammers using compressed air as the working medium were used in place of steam hammers. The smoke from the smiths' hearths was drawn by fans through overhead trunks and exhausted through the roof.
>
> At the forge the raw material was received from the steel works in billets of adequate size for the calibre of shell to be manufactured without undue waste. The steel in the billets was of a higher grade than that used for general engineering purposes, having an ultimate tensile strength of 40 ton/in^2. The billets were delivered by rail and lifted from the trucks by electric overhead travelling cranes – a reflection of H.E. O'Brien's experience at Horwich. Instead of using hooks and slings, each crane was provided with a powerful electric magnet with a capacity to lift at one time five billets suitable for the manufacture of 60-pounder shells.
>
> For storage, the billets were stacked in the space between the railway lines and

---

79 The billets were the rough cast steel bars that were supplied by the foundry.
80 *The Engineer*, Vol.CXXII

the furnaces, from which stacks the cranes lifted them when required at the semi-gas-fired furnaces. In that type of furnace the billets were actually heated by the combustion of the gas from the coal gasified within the furnace and not by the actual combustion of the coal itself. The billets were pushed through the furnace in a continuous process where, on the far side, when white hot, they were withdrawn and quickly transferred to the forging presses. No female labour was employed in the forge as the heat in that working area was considered somewhat trying for them.

The vertical hydraulic forging presses were worked by water pressure at 1,200 lb/in^2. A carriage carrying two moulds moved horizontally between the forging cylinder and the ejecting cylinder thus, while one billet was being forged the preceding one was being ejected. There were two forging punches on each press, which moved at right angles to the carriage carrying the moulds so that on alternate forging strokes the punches were plunged into a water pot to be cooled. Pumps and rams that supplied the pressurised water used by the forging presses were electrically driven, the current to the motors being controlled automatically so that as the demand for pressurised water fluctuated, the number of pumps working could be increased or decreased.

The parts of the presses actually coming into contact with the hot billets required very frequent renewal by tool room staff, particularly in the case of the punches and their components. The tool rooms were staffed by both male and female labour and the replacement parts were produced on the basis of specialised machine working, women carrying out the greater part of this work. The skilled male workers, who were also responsible for supervision of the tool room and instruction of the women, undertook the renewal of the hydraulic valves and other delicate parts of the hydraulic system.

Rolling ways facilitated the passage of forged billets under gravity to tables located at the head of each of the production lines inside the factory. The production lines were laid out in four groups of nine self-contained bays. In each bay, on either side of the central passageway, there was a row of machines, each one of which was different and performed a different operation on the shell forging. Within the factory, work moved in a continuous straight line; a forging, once taken from the rolling way table into a particular bay remained in that bay and on one side of it until, having been passed from one operator to the next, it emerged at the opposite end completely machined. The lines of control of the factory were at right angles to the production lines, all the machines on that line of control were of the same type and performing the same operation. A separate foreman was placed over each such row of machines so that each foreman had one, and only one, class of machine under his charge.

Certain of the machining operations were simpler than others and could be performed more quickly. To prevent congestion, and secure maximum output, the quickest and easiest operations were used to set the production rate. The rate for the more difficult operations was brought into line by providing more than one machine in each bay for that stage of the process and, therefore, in any one bay there were places where two or more similar machines in succession performed identical work. The soap and water cutting fluid was supplied from a central pumping station via overhead delivery pipes with branches to each machine. The used fluid ran off into a concrete gully beneath the machines and was returned to a suds well and pumping station at each end of the workshop, where it was strained and returned to the delivery system for re-use.

Female operators at work on 60-pdr shells inside one of the NPFs, 19th April 1916 - [MUN5-290-002 © TNA, Kew]

To add to the workers' comfort the floors were of wood, and the factories were steam-heated throughout. For this latter purpose, steam pipes were carried on each row of columns at a height of about 10ft above floor level. As the factories were brought into commission, the number of workers was increased to keep pace with the growth. Labour had to be recruited and trained as the construction works progressed but, as the machine tool manufacturers had been able to supply machines that were readily understood and handled, 14 days were generally sufficient to teach a future female operator her work, although in some cases up to a month would be required. The firm in charge of the factory trained all of the people, and the operators themselves, when properly trained, made excellent instructors. At first, one shift of 8 hours was worked, but as more operators became available second and third shifts were started, running from 06:30 on Monday morning to 22:30 on Saturday night, with breaks for meals.

The senior foremen, whose offices occupied a bay in the centre of the factory, were mostly shell machinists before the war at one of the big armament works; the under foremen having been trained for their special work after the factory was started. A small number of trained skilled blacksmiths and tool fitters were employed as well as a few male tool setters who were principally engaged on setting up and keeping the machines and in training female tool setters. The female workers undertook all the actual shell machining and finishing in the NPFs; the success of this workforce being a result of the attention paid to the design of the machines and to the care and patience taken during training. By the time that the NPFs were completely staffed and in full production the number of male employees was less than 5% of the workforce.

View of the 60-pdr and 9.2-in press bay, main bay, and furnace bay at one of the National Projectile Factories, June 1916. - [MUN5-290-003 © TNA, Kew]

CHAPTER NINE National Projectile Factories

So it was that, within just fifteen months, twelve new projectile factories had been designed, constructed and brought into operation, Major O'Brien being able to report on 28th October 1916 that the NPFs had attained their ultimate weekly production, as follows: —

Harper Bean & Son	Dudley	6,000 x 6" H.E.; 15,000 x 60-Pdr Shrapnel
Dick Kerr & Co.	Hackney Marshes	15,000 x 6" H.E.
T. Firth & Sons Ltd	Sheffield	13,000 x 60-Pdr H.E.; 8,000 x 60-Pdr Shrapnel
Beardmore & Co.	Mile End	6,000 x 6" H.E.; 6,000 x 60-Pdr Shrapnel
	Mossend	Forging Plant
	Cardonald	6,000 x 8" H.E.
Vickers Ltd	Lancaster	3,500 x 9.2" H.E.; 6,000 x 6" H.E.; 6,000 x 60-Pdr Shrapnel
Babcock & Wilcox	Renfrew	500 x 12" H.E.; 10,000 x 60-Pdr Shrapnel
Leeds N.P.F.	Hunslet	200 x 15" H.E.; 2,000 x 9.2" H.E.
Cammell Laird & Co.	Nottingham	2,000 x 9.2" H.E.; 6,000 x 6" H.E.
Hadfields Ltd.	Sheffield	4,000 x 9.2" H.E.
Armstrong Whitworth	Birtley	4,000 x 8" H.E.; 6,000 x 6" H.E.; 4,000 x 60-Pdr Shrapnel
G.& J. Weir Ltd.	Cathcart	2,000 x 8" H.E.

Meanwhile, changes had already taken place at the Ministry of Munitions. Lloyd George had been appointed Secretary of State for War on 6th July 1916, following the death of Lord Kitchener, and Fowler had been appointed Superintendent of the Royal Aircraft Factory at Farnborough on 21st September 1916. The mobilisation of the nation's engineering resources to produce adequate munitions of war was complete, and some idea of the enormity of the task that had been undertaken in just fifteen months by Major O'Brien can be gauged from the known costs of capital expenditure on seven of the NPFs:[81]

---

81 National Archives, MUN 4/6738

Factory	Land & buildings	Plant & machinery	Total
Dudley	208,006	579,968	787,974
Hackeny Marshes	173,664	478,887	652,551
Cardonald	278,887	828,760	1,107,647
Lancaster	410,723	1,169,490	1,580,213
Renfrew	84,842	195,632	180,474
Birtley	340,235	552,609	892,844
Cathcart	33,750	2,392	36,142
	£1,530,107	£3,807,738	£5,337,845

With the matter of production sorted out, Lloyd George was about to tackle the question of the effective distribution of munitions and other supplies to the front line. For this he needed key railway personnel, and what better people could he choose than those who had done such sterling work for him at the Ministry of Munitions. As soon as he became Secretary of State for War, Lloyd George sent a request to Sir Douglas Haig that he should invite Sir Eric Geddes over to France to look into the matter of transport. The consequences of this move were to have a direct bearing on the war service of Eoghan O'Brien, but just before he left the Ministry of Munitions sadness was to enter his life again when on 26th October 1916 his mother, Ellen, died at home at *Mount Eagle*.

Light Railways Directorate

We have already noted in Chapter Seven that during the early months of the War the main system for supplying the BEF was the existing network of standard-gauge railways. However, standard-gauge lines took time to build and railheads could not be brought too close to the front line otherwise they would have been subjected to damage from enemy fire. Horse-drawn road vehicles or men and mules were used to bridge the gap of approximately eight miles between the railheads and the front line trenches. The handful of motor vehicles that existed was unreliable and their narrow tyres were unsuitable for the conditions they were expected to work in. Sir Percy Girouard was sent to examine the transport situation in France in late 1914, and his appointment as Inspector General of Transport (IGT) had been considered at the War Office[82], but at that time the matter was dropped. Lord Kitchener was not in favour and in response to a suggestion of using light railways, as the Germans and French were doing, he firmly rejected the idea, stating "that is not our way of working".[83]

Whilst he was still Minister of Munitions, Lloyd George, with the permission of the War Office, had sent Sir Eric Geddes over to France on one occasion to look into matters regarding the recovery and transport of salvage. As no comprehensive plan existed for sustained supply to the troops at the front line, it is hardly surprising that the account of the situation that Sir Eric presented on his return was disquieting. The munitions programme was starting to bear fruit and Lloyd George seriously questioned whether the great increase in output that was anticipated could ever be delivered at the front. He therefore suggested to Lord Kitchener that, with a view to the better organisation of transportation, Sir Eric should be sent to France to conduct an investigation and make a report. However, Lord Kitchener's opinion was that it was purely a military matter, into the sanctity of which no profane civilian could be allowed to intrude. In recalling the events, Lloyd George was later to record that Lord Kitchener "was, by this time, suffering from that growing inertia and ossification of the mind which so gravely impaired his usefulness during his last months of office".[84]

Fate was then to intervene when, on 5th June 1916 Lord Kitchener was drowned after HMS *Hampshire*, the vessel on which he was being carried to Russia, struck a mine in a Force 9 gale and sank. The next day Lloyd George was appointed Secretary of State for War, and transportation was among the first matters he took up. On 7th August he appointed Sir Eric Geddes (whose services had been placed at his disposal by the Minister of Munitions) to investigate transport arrangements

---

82 Henniker, Col. A.M., *The Official History of the Great War*, 1937
83 Davies, W.J.K., *Light Railways of the First World War*, 1967
84 War Memoirs of Lloyd George

in connection with the BEF, both at home and overseas[85]. Following investigation of the U.K. situation, Sir Eric, accompanied by a small staff of two civilian experts and an officer from the Movements Directorate of the War Office, J.G. (later Sir George) Beharrell, Lieut-Col. H. Osborne Mance RE and Brig-Gen. P.A.M. (later Sir Philip) Nash, proceeded to GHQ France on 31st August 1916. Sir Eric brought a fresh mind to bear and looked at the entire technical transportation problem — the carriage of personnel and stores from one place to another. The key question was, what did the army want transported daily, weekly or monthly as the case might be, and between what points? Once these particulars had been obtained, Sir Eric could apply his expertise in determining the best way of doing it, taking existing facilities into account but not tied to them if his experience suggested better ways.

In the course of Sir Eric's investigations the Commander in Chief, Sir Douglas Haig, made a striking remark to him. "Warfare", Sir Douglas had said, "consists of men, munitions and movements — we have got the men and the munitions, but we seem to have forgotten the movement." Accordingly, Sir Eric and his team made a very detailed examination of the tonnage being carried and to be carried in the future, and of the points between which it had to be moved. At the other end of the transportation chain he took the anticipated weekly output of munitions, month by month from August 1916 to July 1917, by which time it was expected to reach its maximum. From the earliest figures it was deduced that the most that could be handled was about 25% more than was being done at the time, while to meet the average requirements of the future a 54% increase in capacity was necessary. Allowing for some weeks to be heavier than others, and also for railway material transportation, an increase of 92% on the existing weekly average was required.

The conclusion was that a drastic overhaul of the methods by which supplies were transported to the front was imperative. Sir Eric's report made it quite clear that an overall organisation was required to plan and co-ordinate military transportation and that a vast increase in transport capacity was necessary. Of critical importance was his recommendation in regard to onward delivery from the standard-gauge railheads to the troops at the front. It was estimated that the amount to be handled might reach 200,000 tons per week and makeshift arrangements could not deal with such a quantity. Neither roads nor standard-gauge railways could be built quickly enough to follow up an advance. The solution that was recommended by Sir Eric was a very rapid development of a proper system of 60-cm gauge tactical light railways in the battle zone to provide transport from the railheads.

The result was the immediate ordering of vast quantities of railway plant and material and the prompt establishment of an organisation to deal with railway transport. Until October 1916 transport forward of the standard-gauge railheads was purely a matter for the Army concerned. The Railway Operating Division

---

85 W.O. Memorandum No.856

(ROD), organised transport up to the railheads where they handed over to the Corps responsible for transhipment and onward movement, but implementation of recommendations contained in the Geddes report was to change all of that. On 18th September 1916 the Army Council approved the appointment of a deputy to the Quarter Master General at the War Office, known as Director General of Military Railways (DGMR), and three days later Sir Eric Geddes accepted the post. As he started to establish his new directorate, the people that Sir Eric selected to assist him were mostly civilian railwaymen and there were two very strong reasons why he followed this course of action. Firstly, because many of the problems to be tackled were ones of which military personnel did not have practical experience on a large scale and, secondly, because he was well acquainted with many railwaymen whom he had already tested while working under him on the NER and at the Ministry of Munitions.

From the outset, three of the five directors were regular officers; one service, Inland Waterways Transport, remaining under its original director, and only the fifth, a new directorate to be managed on commercial lines, namely docks, was under a temporary officer from civilian employment. That officer — Brig-Gen. R. L. (later Sir Ralph) Wedgwood, another senior officer from the NER — already had much experience of military conditions in France as a Deputy Assistant Director of Railway Transport before moving to the Ministry of Munitions. Two members of the team who had accompanied Sir Eric to France during his investigations; J.G Beharrell Esq., and temporary Brig-Gen. H. O. (later Sir Osborne) Mance, were respectively appointed to the positions of Assistant DGMR and Director of Railways, Light Railways & Roads on 25th September 1916.

Almost immediately, Sir Eric was asked by Sir Douglas Haig to reorganise military transportation in France and this proposal by the Commander in Chief was agreed to by the War Office, Sir Eric Geddes being appointed Director General of Transport (DGT) in France on 10th October 1916. On the very same day the third member of his investigating team, Brig-Gen. Philip A.M. Nash, was appointed Deputy DGT and Maj-Gen. Philip G. Twining was appointed Director of Light Railways. Sir Eric had then got his key players in position and for some months thereafter, he occupied the two positions of DGMR at the War Office and DGT in France. It was realised that the work involved with the two directorates would be so great that Sir Eric would need assistance and so Sir W. Guy Granet, the general manager of the Midland Railway, was appointed as his Deputy DGMR, on 20th October 1916. Granet was given specific responsibility for control of the directorate's activities for the home railways.

A vast improvement had to be made in the means of distribution from railheads to the front line destinations. A very large extension in the amount and use of light railways was intended so that the light railway service would be effective. Maj-

Gen. Twining's Light Railways Directorate, which was soon separated from the roads directorate, was responsible, under the DGT, for the existing light railways and for the expansion of the networks in advance of the standard-gauge railheads to meet the requirements of the relevant army commanders. In each of the five Army areas an Assistant Director of Light Railways (ADLR) was appointed to keep in close touch with requirements in his particular Army and who was also responsible for the construction and operation of light railways in his Army area. It was undoubtedly Sir Eric's personal knowledge of Eoghan O'Brien's abilities that was to result in him being the first ADLR to be appointed on 7th November 1916 with the temporary rank of Lieut-Col. whilst so engaged. Other ADLRs, who were appointed during the last two weeks of November, were Lieut-Col. C.W. Myddleton; Lieut-Col. A.G.T. Lefevre; Lieut-Col. A.E. Hodgins; Lieut-Col. H.L Bodwell, DSO; and Lieut-Col. M.H. Logan.

Lieut-Col. H.E. O'Brien was the liaison officer for the Third Army, which was under the command of General Edmund Allenby, and had its field of operation around and to the south of Arras. The light railways were built to a gauge of 60cm with 20-lb/yard flat-bottomed rail laid on steel sleepers on a ballasted formation.[86] Wooden sleepers were also used and, where the ground was soft, they were spaced between the steel sleepers. No makeshift arrangement could have handled the tonnages required; the locomotives and tractors needed the ballasted track. The system had to be treated like a standard-gauge railway to deal with the traffic, but with the added difficulty of more frequent disruptions due to shellfire and bombing from the air. Light railways were developed not as a system of mere tramways, but as a complete railway system managed by railway experts and worked by experienced railway personnel.

Laying of light railway track at Feuchy in 1917 - [©IWM Q05248]

The light railways were far from an economical means of transport, the maintenance during 1917 requiring on average fifteen men per mile, which resulted in costs being about 4½ times that of maintaining standard-gauge track. Except in the Somme and Ancre valley, very few light railways were built before 1917, but after the formation of the Light Railways

---

86 The light railways were often loosely referred to as Décauville lines after the 60cm gauge system, based on portable pre-assembled track panels, that was developed by Paul Décauville. His family owned a beet distillery at Criel, near Paris, and the system was first used in the harvesting of sugar beet. An example was displayed at the Paris Exhibition of 1889 and a substantial business was eventually built up. The term Décauville was probably appropriate for the military lines that were laid in the forward areas, but for the supply lines from the main railheads in the back areas, where steam locomotives were utilised, the track formation was generally of a more substantial nature.

Directorate construction took on a new urgency. During the first six months of 1917 a total of 499 miles of light railway was laid, and light railway repair shops and stores were provided at Berguette. The severe frost, which began in the middle of January 1917 and lasted five or six weeks, impeded the construction of the light railway system but, as progress materialised, the 60-cm gauge light railways played an increasing part in the Army offensives later in the year.

At the time of the Arras offensive in April 1917 the light railway system in the Third Army area was incomplete and its operation for army purposes had only just begun. At that time it did not serve gun positions, but a considerable tonnage was conveyed from the standard-gauge railheads to light railway railheads, where supplies were then transferred to road transport. Large numbers of troops had been trained in the rapid laying of 60-cm gauge track with a view to following-up successive advances, and between 9th April and 1st September 120 miles of light railway extensions were laid in support of fighting requirements. Control boxes were erected at various points along the light railway lines and trains were controlled from one box to another by means of telephone communication. Locomotives were utilised for movement of traffic in the back areas and small tractors for forward area working. The light railways succeeded in delivering large amounts of ammunition in areas that were inaccessible to road transport. By June 1917 light railways in the Third Army were handling an average of approximately 700 tons per day, the average weekly throughput distributed from the standard-gauge railheads eventually equalling 34,000 tons.

Third Army instructions as to the manner in which orders were to be accepted from divisions were based on wagons and motive power available, and priority for various traffics was as follows:

(a) siege and heavy ammunition;

(b) light ammunition;

(c) engineers' materials;

(d) rations;

(e) empty ammunition boxes, cartridge cases, etc., and

(f) other stores such as ordnance stores, salvage, coal, etc.

Wounded men returning from the front were carried in preference to all other traffic.

On 4th May 1917 an ammunition dump took fire at Arras 'Q-Dump' and ammunition exploded for six hours. The personnel of the 31st Light Railway Operating Company worked unflinchingly for two hours clearing wagons, tractors and locomotives from the yard, which were in close proximity to the exploding dump, and this was done at great risk of their lives. As this action under fire coincides with the timing of Lieut-Col. O'Brien's first Mention in Dispatches[87], it is fair to assume that he was involved with his men in this particular rescue of the railway equipment. Notice of his being Mentioned in Dispatches was published in the *London Gazette* dated 15th May 1917. A second explosion took place on 7th May at a small arms ammunition dump, during which a number of British West-Indian soldiers lost their lives and several were wounded and badly burned. The two motor cars of the 31st Company took fire and had to be sent to workshops for repair. The gallantry on the part of the personnel concerned was again particularly noticeable.[88]

At the commencement of 1917 the heavy losses resulting from the activities of enemy submarines stressed the necessity of adopting every possible means to economise in the use of shipping. There was also a need for an alternative route of supply, shorter than the sea journey, for movement of personnel and supplies in support of the allied forces in the near-east theatres of war. A scheme for developing a railway operation over the 1,460 mile route from Cherbourg on the Channel coast of France to Taranto in the south of Italy was approved in principle at a conference held in Rome on 7th January 1917. One week later Sir Guy Calthrop, General Manager of the L&NWR, led a mission to France and Italy for the purpose of investigating the practicalities of the overland railway service. On 7th February this mission presented its report, which decided in favourable terms on the proposed Mediterranean Line of Communication.

On 23rd March 1917 it was decided to proceed with the scheme; construction work being placed under the supervision of Lieut-Col. (later Sir Charles) Morgan, formerly Chief Engineer of the LB&SCR. A great deal of work was involved including the construction of a marshalling yard and transit sheds at Les Flamands, Cherbourg, and a special wharf, transit sheds and sidings adjoining the Mare Piccolo at Taranto. In addition, rest camps were erected at Cherbourg, Saint Germain, Faenza and Taranto, and *halte repas* were provided at twenty locations en-route. The works at Cherbourg were finished and brought into use at the end of August, but those at Taranato, which were on a much larger scale, were not completed until December. Despite the fact that construction works at Taranto only commenced on 15th May, the first passenger train left Cherbourg on 28th June and the first goods train on 8th August 1917.

Prior to the establishment of the Mediterranean Line of Communication a General Head Quarters (GHQ) had been established in Cairo in March 1916 to command the growing Egyptian Expeditionary Force (EEF), but by early 1917 things were

---

87 Mentioned in Despatches — the lowest form of recognition for an act of gallantry announced in the *London Gazette*. Originally there was no award as such; but during World War 1 it was decided that an oakleaf emblem could be worn with the ribbon of the Victory Medal, denoting the mention.
88 National Archives, WO 95/4056

not going well in the Egypt & Palestine theatre of war. Two unsuccessful attempts had been made by the EEF to drive Turkish forces from the Gaza defences; one in March and the other in April. The commander of the EEF, Gen. Sir Archibald Murray, was recalled to London and Gen. Edmund Allenby was selected to take his place from 28th June 1917. Lloyd George charged Gen. Allenby with the express task of capturing Jerusalem from the Turks by Christmas 1917. On arrival in Egypt, the first thing that Gen. Allenby did was to relocate GHQ from its comfortable quarters in a first-class Cairo hotel to Kelab, near Khan Yunis, not far from the front line; an act that helped to boost the morale of the troops.

Before he did anything else, Gen. Allenby had to quickly gain an appreciation of the situation on the ground. In this he was assisted by Lieut-Gen. Sir Philip Chetwode, who considered that well-planned preliminary arrangements were necessary prior to any attack. It was felt that the enemy considered EEF progress would be set by the rate at which standard-gauge railways could be built. With the ground transport then available it was not possible to maintain more than one division at a distance of twenty-five miles beyond a railhead. This was insufficient to keep the enemy on the run after a tactical success and would allow the Turks more than enough time to re-group. Lieut-Gen. Chetwode was of the opinion that suitable and flexible organisation of transport, assembly of material at railheads, training of plate-laying gangs, and use of the earthwork formation of captured railway lines were all very important items in giving the enemy less breathing space.

On the tactical side it was realised that in order to ensure the fall of Jerusalem it was necessary first to break through the Gaza-Beersheba line. The German commander Eric van Falkenhayn, the former army Chief of Staff, who had recently arrived in Palestine, oversaw this Turkish line of defence. It was known that Gaza was the strongest point of defence and that it would require a prolonged artillery bombardment before a ground attack could have a chance of success. Even then, any success would only have been partial, as there was no guarantee of completely breaking the enemy's line. Nor would an attack on the centre of the Turkish line offer any better chance, as it was there that they held the best ground. Both of the earlier attacks at Gaza had to some extent failed due to water shortages — an ever present worry in desert warfare — and Gen. Allenby appreciated that establishing command of water supplies would be a key component in his wider plan of capturing Jerusalem. The ground position around Beersheba was, however, very different and Allenby's strategy centred on securing Beersheba and its water supplies at an early stage.

The Turkish forces at Beersheba were supplied via a single-track 1.05m gauge railway line which connected with the Haifa–Jerusalem line at Junction Station, about eight miles south of Ramleh. At Deir Sineid, some eight miles to the south of Junction Station, a branch line ran off to the south-west to Beit Hanun, from which location the positions around Gaza were supplied. There was a shortage of rolling stock on this railway and the Turks had resorted to hacking down whole olive groves to provide fuel, as there was no coal available for the locomotives.

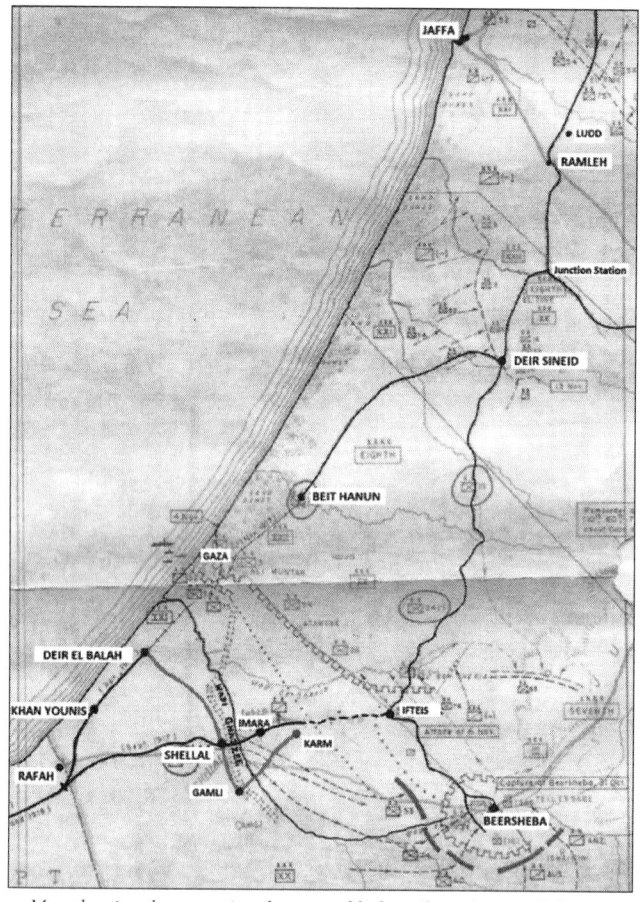

Map showing the operational area and light railway lines in Palestine

Having decided upon his strategy, Gen. Allenby needed to amass reinforcements of men, artillery, shells and tanks, for he was adamant that he would not proceed until he was certain of victory. The Mediterranean Line of Communication was fundamental to the purpose of meeting his requirements for supplies. Construction of the standard gauge single-track line from Kantara, on the east bank of the Suez Canal, had reached El Arish, a distance of 155km, on 21st December 1916 and by the middle of March 1917 the railway had reached Rafah (200km), and was subsequently extended via Khan Younis to a railhead at Deir el Balah (221km). In June 1917, a commission led by Brig-Gen. J.W. Stewart, who with Col. W. McLellan (of Merz & McClellan) and Lieut-Col. L.S. Simpson reported on the railway situation as it affected operations of the EEF. At the time of their visit, a standard-gauge branch line was being constructed from Rafah to Shellal, where an embankment and bridge were being built to cross the Wadi Ghazzee to facilitate an extension north-eastwards to connect with the Turkish line into Beersheba. To meet the increasing needs of the EEF the doubling of the line between Kantara and Rafah was approved to give a capacity of 28 trains per day, of which 16 would run on from Rafah to Deir el Balah and 12 from Rafah to Shellal.

Co-incidental with these events Eoghan O'Brien was promoted to succeed Brig-Gen. G.H. Harrissons as Deputy Director of Light Railways on 25th June 1917 with the temporary rank of Colonel; his place as ADLR being taken by Lieut-Col. H.M. Stobart. At the time of his appointment it had already been decided to construct a 2ft 6in gauge light railway from the railhead at Deir el Balah towards the Wadi Ghazzee and from that point along the line of the wadi to Shellal, 12 miles to the east. Gen. Allenby's first-hand experience of Col. O'Brien's work with the Third Army was probably the reason that, in addition to his overall responsibility for light railways in France, he was assigned special duties for the supply of railway troops, equipment and rolling stock for tactical light railways that would be required in support of the Palestine campaign.

Col. O'Brien was in Italy until mid-October ensuring that all men and materials required were flowing smoothly down the Mediterranean Line of Communication in a timely manner. His fluency in Italian and French would, once again, have been of great help in getting issues sorted out with local officials. The light railway to Shellal was constructed under the direction of the ROD from July 1917 onwards, and this 12 mile line enabled ammunition, supplies, and stores to be taken directly up to the right flank at Shellal where large quantities of railway material (much of it recovered from redundant light railway lines in the canal defence zone) were piled up in readiness for pushing out the railheads immediately the advance began. Shellal was also a position of great strategic importance with an invaluable water supply.

About the middle of August it was decided that the attack on the Turkish front line in southern Palestine would begin in September. However, it was soon realised that a September advance was not absolutely necessary and Gen. Allenby decided that instead of making an early attack it would be far more beneficial to wait until his Army was fully prepared. To enable about two-thirds of the force to carry on a moving battle while the remainder kept the enemy pinned down, it was necessary to reinforce strongly the transport services. It had originally been suggested that motor transport might be used as a means of distribution from the railheads, but it was recognised that this form of transport could not be relied upon on a wide scale.

Gen. Allenby therefore agreed to allow the extension of the standard-gauge railway east of Shellal to be begun sooner than he had originally planned. It was imperative that this railway construction should not give the enemy any indication of the

British intentions, but Allenby was willing to take the risk in order to pre-empt difficulties in supplying his front line. He ordered that this extension, to a railhead north-east of Karm, should be completed on the evening of the third day before the attack. Under the direction of Major Jordan-Bell, the standard-gauge line was first extended to Imara, where a station was opened on 28th October; and then pushed on to a point ¾ mile north-north-east of Karm, which was reached on 3rd November.

Regarding the 60-cm gauge light railways, it was decided that a line out towards Gamli from Shellal would not be started before the sixth day prior to the attack, but it was to reach Karm by the day preceding the opening of the fighting at Beersheba. A new light railway line was to be started at Karm when fighting had begun, and carried nearly three miles in the Beersheba direction early on the following morning. From his experience with the Third Army at Arras, Gen. Allenby knew that Col. O'Brien was the right man to get these light railways constructed on time and had him assigned to Palestine for 'special duties'. The line from Karm to Khasif was laid with astonishing rapidity by working parties of Sikh Pioneers, and all of the new lines, although of short length, were of immense value in supplying the front.

Most of the light railway line along the Wadi Ghazzee was under enemy observation and within range of their guns. When it was in running order trains had to run the gauntlet of artillery-fire on this section on bright moonlit nights, but the line seemed to live a charmed life as no material damage was sustained. Despite these difficulties it worked in an orderly and efficient fashion, and on one occasion carried as much as 850 tons of ammunition to the batteries in one day. In naming their tiny locomotives the drivers showed their strong appreciation for their comrades of the sea, the *Lion* and the *Iron Duke* were always tuned up to haul a maximum load, but the pride of the yard was one engine that carried the name *Jerusalem Cuckoo*.

With the attack on the Turkish front line in Palestine imminent, Col. O'Brien had sailed from Taranto on board the HT *Aragon*, disembarking at Alexandria on 20th October. A few days after his departure from Italy, the Italian Army met with disaster when, on 24th October 1917, the Austrians, assisted by German divisions, attacked on a wide front at Caporetto. The Italian 2nd Army broke, failing to hold the line of the Tagliamento, and finally came to a halt behind the Piave. The result of this was that all the efforts of the French and British were diverted to reinforce the Italian line of defence and the services to Egypt and Salonica via the overland railway route had to be temporarily suspended.

Col. O'Brien arrived at the GHQ of the 1st Echelon of the EEF on 28th October. GHQ was situated near the railway junction at Rafah, with the standard-gauge lines from Rafah to Deir el Balah and Shellal on either side of the encampment. Heavily laden trains ran day and night with a mass of stores and supplies. The

CHAPTER TEN Light Railways Directorate

heaviest trains were run at night, and the returning empties were moved at a speed suggesting the urgency of clearing the single-track lines for a fully loaded train waiting at Rafah for the signal to proceed with its valuable load to a railhead. Enormously long trains, most of them hauled by former London & South Western Railway '395' class 0-6-0 goods locomotives, bore munitions, food for men and animals, water, medical supplies, guns, wagons, caterpillar tractors, motor cars, and other equipment required for the largest army that had ever operated in Palestine. Good control, not only on the railway system, but also in the forward supply yards, prevented congestion. When a train arrived at a railhead it was split up into several parts and well-drilled gangs of troops were allocated to each wagon. One contemporary report stated that "the goods were cleared away from the vicinity of the line with a celerity which a goods yard foreman at home would have applauded as the smartest work he had ever seen".[89]

Preparatory movements for the attack began on the night of 20th/21st October; initially only a force sufficient to form supply dumps and store water being sent forward. By 31st October Beersheba had been taken, but in the meantime important events were happening at the other end of the front line. The preliminary bombardment of Gaza had begun on 27th October and on the night of 1st/2nd November the attack on the city commenced. One spectacular incident occurred during the bombardment of Gaza, in which British and French naval ships joined with army artillery. Taking a signal from a seaplane, HMS *Raglan* managed to score a direct hit on a Turkish ammunition train at Beit Hanun, the railhead north of Gaza. The whole train blew up and fragments of the train were scattered over an area of several hundred square yards, resulting in an extraordinary scene of torn and twisted railway wreckage. Following the capture of Gaza on 4th November Col. O'Brien set up a light railway organisation at Beit Hanun

The autumn offensive continued with infantry divisions being given the objective of securing the important Junction Station on the Turks' Haifa—Jerusalem railway line on the 13th November. A big step forward was made in the early afternoon by over-coming stubborn resistance at Mesmiyeh, but difficult terrain slowed the advance before a charge by 800 cavalry troops — assisted by artillery and machine gunnery — cleared El Mughar by the evening of that day. Consequently, the infantry divisions launched an attack on a ridge north-west of Junction Station, and an advance began during the night, but it was strongly counter-attacked and had to halt till the next morning. At dawn, having secured the best positions on the rolling hills to the west, armoured cars entered Junction Station, and by 7.30 a.m. it was occupied. The capture of this important railway junction succeeded in cutting off communication between the Turkish 7th and 8th Armies.

Junction Station was no sooner occupied than light-railway staff under Col.

[89] Massey, W.T., *How Jerusalem was Won*, London 1919

O'Brien were brought up from Beit Hanun. The line formed part of the Turkish 1.05m (3ft 5¼in.) gauge network, but the British had no suitable rolling stock to fit it. Luckily, two locomotives and forty-five wagons were found intact, as well as two large guns on trucks. This equipment was of immense value to the EEF, and to a large extent it solved the transport problem, which at that moment was a very anxious one indeed. The enemy continued to shell the station during the morning, trying in vain to damage this rolling stock, but A/Sgt. W.G. Clarke, assisted by Cpl. F.R. Owens steamed one of the two captured Turkish locomotives under most difficult circumstances. The water tanks had been destroyed, as a result of which the two men spent hours in filling up the locomotive using a water jug and basin found in the station building. Meanwhile, L/Cpl. L.A. Cook was busy examining the rolling stock. These men made it possible for Sapper J. Warlow to take the first locomotive out. The Turks must have been mortified later in the morning to see the locomotive steam out of the station and into a cutting, which provided effective cover from their field-gun fire. Of course, the light railway staff were delighted at their success, and the trains that they soon had running over the system were indeed of great value to the troops at the front. It was for this particular gallant action under fire on 14th November 1917 that Col. O'Brien was Mentioned in Dispatches for the second time; notice of the event being published in the *London Gazette* dated 11th December 1917.

Junction station, Bersheeba, after its capture — one of the Turkish 1050mm gauge locomotives is just visible at the right. - [©IWM Q13165]

On the morning of 15th November Gen. Cox's New Zealand mounted rifles brigade seized Ramleh and then advanced to capture Ludd, and the New Zealanders were in Jaffa by noon on 16th November. The next task that A/Sgt. Clarke and his men turned their attention to was the recovery of a third Turkish locomotive, which lay on its side at the bottom of the embankment at Deir Sineid. They worked 72 hours continuously to re-rail it and two of the Turkish locomotives were ready to operate by 19th October when the first train movement took place to Beit Hanun using captured rolling stock. However, due to the lack of water no useful train running was possible until the night of 21st November, and on the next day the first supply train reached Junction Station. All of the Turkish locomotives had failed by 25th November, and they were not used thereafter as four locomotives from the 3ft 6in. gauge Luxor—Assouan section of Egyptian State Railways had arrived on that day together with a number of wagons. Later two more locomotives were sent down from Egypt, but although the new motive power was available, difficulties with the water supply and problems with telephonic communication, together with malicious damage to the line, disrupted the service. Water was not available between Deir Sineid and Lineh till 1st December.

The whole of the line from Junction Station to Deir Sineid was not in running order, but broken culverts requiring only minor repairs were given immediate attention. A deviation necessitated by a destroyed bridge at Junction Station, and the relaying of 3km of destroyed track, was completed on 3rd December and the line open to Ramleh and Ludd. During this period, preliminary steps were also taken towards repairing the destroyed bridges on the Jerusalem line at km57 and km58. On 1st December a train was run through to Artrif, near Deir Aban station on the Jerusalem line. In the whole period of 17 days from 21st November to 7th December, 53 Up and 35 Down trains were run and the total carried amounted to 4,391 tons of supplies and 390 tons of ammunition, as well as 1,634 troops; 3,543 wounded; 353 natives; and 795 prisoners. Two specials were also run; one for a Corps of H.Q. staff; and the other carrying Construction Company Staff and baggage.

With the rapid rate of advance, it became necessary for Col. O'Brien to obtain railway troops to operate the 1.05m gauge network captured from the Turks. This had become a vital transport artery for the EEF and, although it was planned to convert it all to standard gauge, there was an immediate need to keep it operational. The 96th Light Railway Operating Company and the 98th Light Railway Train Crew Company were sent out from England in early December for this purpose. The 1.05m gauge lines were converted to standard gauge during December 1917 and January 1918 and when that work was completed they came under the control of the ROD. This released light railway troops who were then deployed on the construction of 60-cm lines from Ludd to Jaffa, and from the latter point northward along the coast to the front line.

When it became obvious that Jerusalem could not be secured without adoption of a deliberate method of attack, there were many matters that required attention. Men and guns were available, but to get them into the line and to keep them supplied was a problem of considerable magnitude. The rains had begun and after each period of rain the sodden state of the ground impaired all movement. From Gaza and Deir Sineid there were only roads available to bring up supplies, and the light soil had become hopelessly cut up during the rains. The main line of railway was not opened to Mejdel until 8th December, and the 1.05m gauge line from Junction Station to Deir Sineid had a maximum capacity of one hundred tons of ordnance stores per day, and these had to be moved forward again by road.

By 2nd December the efforts of the troops working on the line of communications had enabled the plan for the capture of Jerusalem to be made. From 4th to 7th December a redeployment of troops took place to enable a concentration of the divisions entrusted with the attack on the Jerusalem defences. The rains brought water-borne problems too, and Col. O'Brien fell ill on 3rd December with diarrhoea, which resulted in him being detained in hospital for three days, but he was fit again by 8th December when the attack on Jerusalem commenced. The action there was over quickly and at 12:30 on 9th December the city was surrendered.

After vacating Jerusalem the Turks took up strong positions on the hills to the north and north-east of the city from which they had to be driven before Jerusalem was secure from counter attack. The Germans prevailed on their ally to make an attempt at retaking Jerusalem, and thirteen powerful attacks were made upon the allied lines on 27th December. However, this venture had disastrous consequences for the Turks and, instead of reaching Jerusalem, they had to yield seven miles of territory, which gave the allies four strong lines for the defence of the city. By 30th December sufficient ground had been secured to provide Jerusalem with an impregnable defence at a cost much less than could have been expected if the Turks had remained on the defensive. Conversion of the former Turkish railway to standard gauge was still a long way from reaching Ramleh and the railway construction parties had to fight against bad weather and washouts. The line from Junction Station to Jerusalem was also in bad order, with a number of bridges down, and it took several weeks to complete its rehabilitation.

With his task in Palestine complete, Col. O'Brien set sail for Italy on 4th January 1918 on board HT *Karoa* bound for Taranto. His service in the field during 1917 had not gone unnoticed and he would probably just have received the news that King George V had approved the award of the Distinguished Service Order to him in the New Year's honours of 1st January 1918. His work at the Light Railways Directorate was not over, and on his return to Italy he had to address another pressing problem. Apart from a few officers required for liaison duties, no standard gauge operating personnel had been sent out from England; the Italian railway authorities undertook the working and maintenance of the British troop train services in Italy under the co-ordination of the Railway Transport Establishment. The Italian authorities were unable, however, to provide any suitable operating personnel for light railways. On 8th January 1918, Col. O'Brien recommended that a Light Railways Headquarters

Staff together with one Light Railway Operating Company should be sent out to Italy together with a supply of light railway track and a due proportion of rolling stock, etc. The 109th Light Railway Operating Company, which embarked for Italy on 18th April 1918, were eventually utilised for working British traffic over the 60-cm gauge Thienne-Calvene light railway.

In preparation for an anticipated German Spring offensive, a fresh examination was made in January 1918 of the transportation system on the Western front. As almost 80% of railway movements from the north to the south and east passed through Amiens, it was an obvious objective of any German advance. In order to connect up the light railway system of all of the allied Armies, it was decided to construct a lateral light railway about three miles behind the front trench system. Col. O'Brien was, therefore, back in France to oversee the construction of that network, which was completed by the middle of March. On 21st March the long-awaited German offensive opened up on the front line of the 3rd and 5th Armies. The attack was so intensive that the 5th Army had to make a prolonged retreat, which continued until 4th April.

The sustained action by the 5th Army during its retreat resulted in German forces suffering a great number of casualties and they became so exhausted that they were never able to reach their strategic objective of Amiens. However, the German advance into the area of the 5th Army was so rapid that considerable effort had to be directed towards the withdrawal of rolling stock. The penetration into the 3rd Army area was not so deep or rapid, and the light railways continued to operate for a time before withdrawal of rolling stock became necessary. The lateral light railway system played a very important role in handling the large number of strategic movements that had to be undertaken. Once it had been established that the German advance had been halted, Col. O'Brien left the field of action. He was back in England by the end of April 1918 and about to be assigned to another important task.

Assigned to two Ministries

The formation of the Ministry of Munitions on the 9th June 1915 has already been referred to in Chapter Eight. Lloyd George and Christopher Addison had to work as quickly as possible in setting up the new organisation and the task of initiating the Contracts Department was assigned to Sir Hardman Lever[90]. Up until September 1917 the organisation of the Contracts Department consisted of the Director General of Munitions Contracts and two Deputy Director General (one for contracts placed at home and the other for contracts placed in the USA). There were ten Directors of branches, seven Deputy Directors, fifteen Assistant Directors and seven Deputy Assistant Directors.

The Contract Costs branch was divided into two sections: (a) accountancy costs, and (b) technical costs. Staff in the accountancy costs section consisted of professional accountants and assistants having special accountancy knowledge. The technical costs section was composed of engineers skilled in rate fixing and estimation of work. Until the beginning of 1918 the branch was a comparatively small one, the staff up to that date consisting of 13 accountants, 15 cost engineers and 6 clerks; a total of 34.

In September 1917 the Contracts Department was taken over by Mr J. (later Sir John) Mann as Controller of Munitions Contracts. After careful consideration, a decision was taken to re-organise the Department into nine branches, viz. seven purchasing branches, a branch dealing with capital and expenditure, and the costs branch; a contracts claims branch was added subsequently. The officers in charge of these branches were graded as Assistant Controllers. Having regard to the increasing variety of types of munitions stores purchased, and the necessity for the strictest economy, it was decided to considerably strengthen the cost branch. The areas of responsibility of the Contracts Department which were clearly laid down in February 1918 were as follows:

(i) The Department is responsible for the price and terms of all contracts;

(ii) The Department has power to conclude all home contracts;

(iii) Contracts in Canada are concluded by the Imperial Munitions Board; and

(iv) Contracts in the United States are concluded by the British Mission in the USA.[91]

It was to this particular department of the Ministry of Munitions that H.E. O'Brien was appointed on 1st May 1918 as Director of Contract Costs at the grade of

---

90 National Archives (PRO), MUN 111/500/38
91 Office Memorandum No.61 of 22nd February 1918

Assistant Controller, reverting to the rank of Lieut-Col.[92] A few days later, on 6th May 1918, Sir Worthington Evans, the Financial Secretary of the Department, submitted a scheme for decentralisation involving, *inter alia*, the amalgamation of Contracts and Supply departments. However, this suggestion was not put into effect and by the end of the War the staff of the Contracts Costs branch had grown to a total of 118, including 32 accountants, 57 engineers and 28 clerks. At the beginning of 1918 a weekly total of 15 accountancy investigations and 16 technical investigations were being carried out. During the year this rose steadily and, by the end of the War on 11th November 1918, Lieut-Col. O'Brien was responsible for a total of 35 accountancy and 142 technical investigations weekly. In addition to these investigations a number of audits were carried out on contracts and Lieut-Col. O'Brien served as a member of the Munitions Contracts Board. Some idea of the overall task that the Contracts Department was responsible for can be gauged from the following table, which covers the period during which he was Director of Contract Costs.

Period	Average number of Contracts (weekly)	Average Weekly Value (£)	Staff
March – June 1918	1,824	Projectiles: 5,330,511 Aircraft: 3,238,720 Mech. Warfare: 1,102,660 Other: 2,669,957 **Total: 12,341,848**	949 – 1089
June - September 1918	1,874	Projectiles: 5,777,045 Aircraft: 3,880,714 Mech. Warfare: 883,410 Other: 3,161,482 **Total: 13,702,651**	1089 – 1244
September – 11th November 1918	2,191	Projectiles: 9,360,736 Aircraft: 3,658,947 Mech. Warfare: 2,987,588 Other: 4,634,039 **Total: 20,641,310**	1244 - 1269

Another important event following Lieut-Col. H.E. O'Brien's return to the Ministry of Munitions in May 1918 was his appointment, together with James Hamer Crompton (Honorary Treasurer of Horwich Mechanics' Institute), as a trustee of the Fielden Horwich Gymnasium Endowment Fund.

By July 1918 it was recognised that much of the vast resource of Woolwich

---

92 About ten years after the War, it was agreed that persons who had served as temporary officers were entitled to use the highest rank that they had attained as a form of address.

Arsenal would soon become surplus to requirements. Sir Vincent Raven, Chief Superintendent of Ordnance Factories, had remained in control until June 1917 and G.H. Roberts, another of the bright engineers from the Horwich School, was Superintendent of the Mechanical Engineering Department. In view of the changing circumstances, the Minister of Munitions appointed the Rt.Hon. T. McKinnon Wood MP to chair a committee to "enquire into and report upon the control, administration, organisation, lay-out, and equipment of the Royal Ordnance Factory at Woolwich... ...and to advise the Minister what, if any, changes are required". The Committee reported as follows:

1st Interim Report	On Danger	6th November 1918
2nd Interim Report	Organisation and Administration	22nd November 1918
3rd Interim Report	Costing System	11th February 1919
Final Report	Confirmation	13th March 1919

Although Lieut-Col. O'Brien would undoubtedly have been involved, particularly in regard to aspects of the 3rd Interim report, there is no direct evidence to suggest that he had any influence in suggesting that the surplus capacity of Woolwich might be used for the construction of railway locomotives. There are, however, a number of significant clues that suggest that he might well have had some part to play in such a decision. Firstly, he served on the Committee on Demobilisation and Reconstruction that had been set up by Council of Government to consider the effects of terminating ammunition contracts. He was so concerned with some of the proposals that were being developed that he submitted his observations on the discussions that had taken place at the 35th meeting of the Committee held on 10th July 1918. In his letter he commented: —

> I think I should point out to the Committee, that after more careful consideration of Mr Stevenson's proposal as to the method to be adopted in demobilising gun ammunition factories I totally disagree with his proposals; the conditions that will arise in regard to the labour situation must be the paramount and governing factors to be considered in coming to a decision on the method to be adopted; such employers as I have spoken to on the subject are unanimous in their opinion that some steps must be taken to keep the labour employed in the shops as far as possible during the transition period from war to peace conditions.
>
> The termination of hostilities will probably result at first in an immediate reaction of feeling which will result in all work being suspended for a week or so; at the end of this period, unless steps are taken to provide employment of some kind, serious trouble will ensue; by working out to completion the work in progress at the factories on day-work rates and reduced shifts a breathing period will be given during which staff will be most urgently required to enable the scheme of subsistence allowances to be brought into force and the many queries and problems that will arise to be dealt with.
>
> It is improbable that the workpeople will refuse to continue working on destructive munitions; on the other hand it is probable that they will be only too glad to have work to do at all; as stated above, few employers will care for the idea of their employees remaining idle for a period of weeks on a subsistence allowance paid by the Government, the conditions too closely resemble those prevailing during a

> strike, and the whole position is more demoralising to the work people than even that of working on useless destructive munitions.
>
> The working out of the gun ammunition to a completed form will not result in any appreciable amount of material being wasted; if this ammunition is not required it will be scrap but the amount of scrap will be the same though in a different form to that which it would have if manufacture was suddenly arrested; at the worst the loss of useful material would be no more than that of a small amount of brass rod used in component manufacture.
>
> I agree that Mr Stevenson's and Mr Collinson's procedure is feasible, but I do not agree that it is expedient, and I am certain that it will give rise to an enormous amount of correspondence and result in many claims; if this procedure is to be adopted I would urge that it should be actually tested in a few small sample factories if only for the purpose of ascertaining what difficulties will arise and to enable an estimate to be made of the time and labour necessary to prepare the schedules and arrive at the payments to be made.
>
> I wish to point out that it is not the duty of the Contracts Department to prepare these schedules, nor have they the staff available for the purpose if it were so. It is the duty of the Supply Department to prepare the schedules and the duty of the Contracts Department to fix the prices at the definite stages selected; this in reference to the first paragraph on page 7.
>
> In connection with the small committees agreed to by the Committee for the purpose of drawing up recommendations for demobilisation of each class of munitions, may I venture to suggest that these committees should be instructed to invite the views of representative manufacturers on the subject; the manufacturers will be the principal sufferers in the chaos and confusion which may ensue on the termination of hostilities and their knowledge of local conditions and of the attitude of mind of labour will be of the greatest value in coming to a decision on the best procedure to be adopted. [93]

From the foregoing it is obvious that Col. O'Brien had a great concern for the welfare of the workers that would be affected. This was something that had always been at the heart of the Horwich tradition established by Aspinall and evidenced in the facilities provided by the L&YR for its staff.

The second point to consider in connection with any involvement that H.E. O'Brien might have had in influencing the Woolwich venture into locomotive construction was his family connection with R.E.L. Maunsell, Chief Mechanical Engineer of the South Eastern & Chatham Railway. The decision to adopt Maunsell's 'N' Class 2-6-0, rather than the proposal of the Association of Railway Locomotive Engineers as the locomotive for mass production at Woolwich, is possibly another relevant clue. The third point worth considering is the presence at Woolwich of a former L&YR colleague, G.H. Roberts, who in 1918 had been promoted to the post of Chief Mechanical Engineer at the Royal Arsenal. Finally, there was H.E. O'Brien's long-standing friendship with W.H. Morton, Locomotive Engineer of the MGWR. Did that particular connection provide the channel for the necessary information to flow to the Broadstone to support that company's decision in 1923 to purchase sets of parts from Woolwich for twelve locomotives? Just after his appointment as Deputy CME of the GSR in January 1925, it was Morton who was

---

93 Serial No.149, 20th July 1918

also to convince his new Board to purchase a further fifteen sets of Woolwich parts.

Although, in the absence of any hard evidence, all of this is speculative; we shall see in a later chapter that Eoghan O'Brien was as effective in behind the scene activities as he was in dealing with matters front stage. The Ministry of Munitions had ceased to have responsibility for the Royal Arsenal at Woolwich on 1st June 1920. Suffice to say that the short sojourn into locomotive construction at Woolwich, although not proving to be an entirely successful venture, did satisfy Eoghan's objective "that some steps must be taken to keep the labour employed in the shops as far as possible during the transition period from war to peace conditions".

Official photograph of the first Woolwich mogul assembled by the MGWR posed at the Broadstone in the first quarter of 1925. It was to have been MGWR No.49, but is shown here bearing its originally assigned GSR number 410 (it was re-numbered 372 in April 1925). - [Green Studio collection, courtesy Sean Kennedy]

With the signing of the Armistice on 11th November 1918 circumstances were about to change for many people. For Eoghan O'Brien, at least, the fact that his position as Assistant Chief Mechanical Engineer with the L&YR was open for him to return to presented a degree of security that was not available to many who were demobilised. The year 1919 began with the good news of his transfer on 28th January to the class of Member of the I.C.E., but on 15th February he had to take sick leave from the Ministry of Munitions. The sudden relief from the pressure of work during the war years might have been a factor that affected his health, but the circumstances and duration of his illness are not known. It appears that he did not return to the Ministry of Munitions before his release from military service on 25th April.

Meanwhile, the General Election of December 1918 had resulted in a landslide victory for Sinn Féin who won 73 of the 105 Irish seats and the Unionist parties 26 seats. The Sinn Féin candidates who had been elected declined to take their seats at Westminster and on 21st January 1919 they established an independent legislature in Dublin called Dáil Éireann. It was against this backdrop that the Irish Centre Party was established in the same month. The party membership was predominantly professional men and women from moderate, middle-class Dublin families. Although it raised some interest it failed to gather sufficient members to make it a prominent force in Irish politics.

The Irish constitutional debate had fundamentally altered since the Edwardian period and the Centre Party espoused a Federal solution for Ireland to reconcile its internal divisions and provide for efficiency in government. In a short foray into the sphere of politics, Lieut-Col. Eoghan O'Brien described the party's vision in an article published in the *Overseas Magazine*.[94] The party advocated the creation of a self-governing state with Dominion status within the British Empire similar to Canada and New Zealand.[95] Provincial assemblies were proposed for the four provinces of Ireland to handle local affairs, and a National Parliament in Dublin would deal with affairs of state. The party argued that this arrangement would allow the interests of both Protestant and Roman Catholic communities to be fairly represented, while avoiding a division of Ireland.

The concept of a federal Ireland had limited appeal to nationalists, especially those in Sinn Féin who feared that it would undermine the political and territorial integrity of an Irish state. For unionists, the federalist solution proposed the dissolution of the Union with Great Britain, which they strenuously opposed. The poor level of support for the Irish Centre Party led it to merge with the dominion movement to form the Irish Dominion League, a small political party which only survived until 1921.

The Westminster Government was also turning its attention to post-war matters. The Ministry of Transport Act was passed on 18th August 1919 and Sir Eric Geddes, who, in view of his public duties, had severed his connections with the NER in February of that year, was appointed as the Minister of Transport. Sir John Aspinall was engaged as Consulting Mechanical Engineer and Col. Lightly Simpson was appointed Chief Mechanical Engineer at the new Ministry. One of the most important issues that Col. Simpson's department sought to address was the question of harmonising standards so as to achieve a wider interchange of locomotives and rolling stock. Whilst under the control of the Railway Executive Committee during the War years, Britain's railways had been operated as a national undertaking. This, more than anything, had revealed the diversity of standards that applied; each railway generally having been concerned only with the constraints that applied within their respective sphere of operations.

The policy of the Ministry in regard to electrification of the railways was to encourage the different railway companies to spend their own capital on electrification projects wherever a well-considered scheme could be produced, especially if it

---

94 *Overseas Magazine*, Vol IV, No.41, June 1919

95 It should be noted that the proposal for Dominion status pre-dated by 30 years the establishment of the British Commonwealth (now the Commonwealth of Nations) in which former territories of the British Empire are recognised as "free and equal" member states under the terms of the London Declaration of 1949.

was concerned with suburban traffic. In March 1920 the Government appointed a Committee, under the chairmanship of Sir Alexander B. Kennedy, to consider the technical points in connection with interchange of electric rolling stock and to advise on any questions that might arise in respect of through running of electric services between contiguous networks. Although the Electrification of Railways Advisory Committee was formed of well-known technical engineers experts from the various railway companies, such as H.W.H Richards of the LB&SCR, Herbert Jones of the L&SWR and H.E. O'Brien of the L&YR, were called to give evidence before the Committee

A provincial branch of the Ministry was established in Dublin, and its structure was described in a report of 30th September 1920. The Director General, Henry G. Burgess was responsible for development and political work, and under him there were two divisions; finance and statistics under Joseph Ingram, and traffic matters under Percy Wharton.[96] H.G.Burgess, who was the L&NWR Traffic Manager in Ireland and also Deputy Chairman of the D&SER, was appointed Director General of the Irish branch in July 1919, and in this position he reported directly to the Minister for Transport.

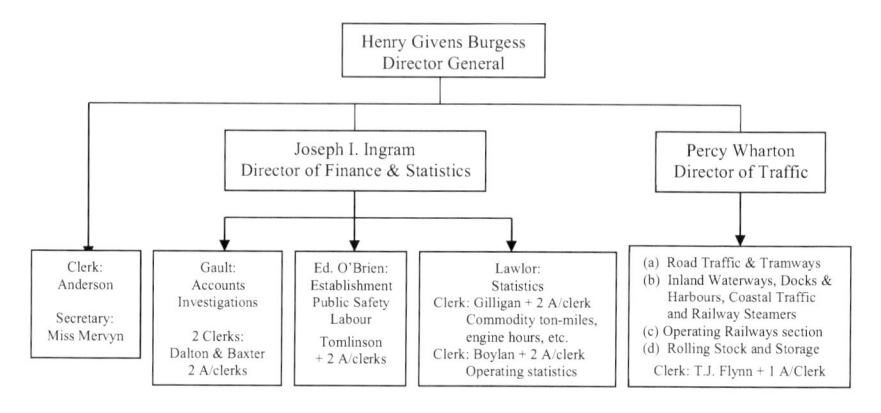

```
                    Henry Givens Burgess
                    Director General
        ┌─────────────────┴─────────────────┐
   Joseph I. Ingram                     Percy Wharton
   Director of Finance & Statistics     Director of Traffic
```

| Clerk: Anderson Secretary: Miss Mervyn | Gault: Accounts Investigations 2 Clerks: Dalton & Baxter 2 A/clerks | Ed. O'Brien: Establishment Public Safety Labour Tomlinson + 2 A/clerks | Lawlor: Statistics Clerk: Gilligan + 2 A/clerk Commodity ton-miles, engine hours, etc. Clerk: Boylan + 2 A/clerk Operating statistics | (a) Road Traffic & Tramways (b) Inland Waterways, Docks & Harbours, Coastal Traffic and Railway Steamers (c) Operating Railways section (d) Rolling Stock and Storage Clerk: T.J. Flynn + 1 A/Clerk |

Organisation of the Irish Branch of the Ministry of Transport

The appointment of the two Directors by the Ministry took effect on 26th September 1919. Joseph Ingram was formerly Secretary of the Irish Railway Clearing House and Percy Wharton had been Traffic Manager on the GS&WR. Percy Wharton's chief clerk, T.J. Flynn, was later to become Director of the Transport & Marine Branch at the Department of Industry & Commerce in the Free State Government, and he was later to serve on the 1938 Transport Tribunal, which was chaired by Joseph Ingram.

As the Irish branch got into its stride the need for specialist staff became necessary.[97]

---

96 National Archives (PRO), MT45/12

97 By December 1920 the staff of the branch had grown to a total of 41 with a budget of £16,138 for the following year, but by that time H.E. O'Brien had returned to Horwich.

CHAPTER ELEVEN **Assigned to two Ministries**

In December 1919 Aspinall received an application from the Ministry for the loan of H.E. O'Brien, who had been elected a Member of the I.Loco.E. on 29th October. It was suggested that he should go over to Ireland with a view to advising Burgess on mechanical engineering matters. Aspinall advised the L&YR Board that it was O'Brien's desire to eventually return to his home in Ireland and that the temporary loan might very possibly lead to his taking a permanent appointment under the Ministry. Eoghan O'Brien departed from Horwich on 23rd December in order to spend Christmas at *Mount Eagle* before taking up his new role as Chief Mechanical Engineer at the Irish branch of the Ministry of Transport, and once again it was G.N. Shawcross who took over as acting Works Manager.

Henry Givens Burgess
Irish Traffic Manager, L&NWR (1898-1919);
Deputy Chairman D&SER (1907-1919);
Director General, Irish Branch MoT (1919-20);
General Manager, LMS (1924-27) - [Author's collection]

Oriel House, *Dublin, where the Irish branch of the Ministry of Transport was based - [Author]*

On 5th January 1920 H.E. O'Brien read his paper, *The application of the Electric Locomotive to Main-line Traction on Railways*, to the Preston Sub-centre of the I.E.E. and repeated it at the Bristol Centre on 7th February. A third reading took place at the Liverpool Sub-Centre on 15th March and just three days later he was transferred to the class of Member of the I.E.E. The paper introduced the subject with a brief historical background of the development of electric locomotives. The author then proceeded to look at the general problems of design, paying particular attention to the weight of traction motors and the effect of their mounting on centre of gravity and height of wheels and bogies. Various curves were used to show the performance of typical electric locomotives compared against steam types, and following sections of the paper dealt with operations and maintenance, and the financial aspects. Tables illustrating the costs of maintenance and repairs, mileage between overhauls, and maintenance periodicities of L&YR electric stock, were included in support of the benefits of electric traction described in the paper.

One of the most interesting contributions to the discussion was that by S.B. Haslam, who had carried out exhaustive inquiries with Tom Hurry Riches into the question of applying electric traction on a six-mile section of the Taff Vale Railway, which he noted culminated in a paper read to the I.Mech.E. at their summer meeting held at Cardiff in August 1906. He was interested to learn from H.E. O'Brien's paper that the difficulties were practically the same as seventeen years earlier, and that they were railway operating difficulties rather than electrical ones.

On 12th February 1920, H.E. O'Brien wrote to the GS&WR drawing attention to the importance of using any possible supply of Irish coal for locomotive purposes. He pointed out that the total locomotive coal consumption on the Irish railways was about 500,000 tons per annum and that this quantity cost approximately £1.25 million at the prices then ruling for imported coal. He stressed the point that Irish coal had never been properly tested on the railways — any experiments had been with small lots of coal and without any attempt to adapt the firebars, brick arches, blastpipes, etc. Also, the engine crews had not been given time to get accustomed to the different coal. In this respect he noted that even on a large English railway a radical change of coal would produce a crop of complaints and failures until such time as the drivers had become familiar with the new conditions

In H.E. O'Brien's opinion previous experiments were valueless, and he felt that any future tests should be made on a lot of 500 to 1,000 tons of coal which should be collected at one locomotive shed, tried first on light traffic and then on heavier duties. He felt that the Locomotive Superintendent concerned should have a free hand to make small modifications to the locomotive under test. However, he was not dogmatic in his views, noting that it was quite possible that the Irish coal might not prove suitable for locomotive purposes; its use could result in unduly heavy coal consumption, or undue wear and tear of boiler. On the other hand, he considered that it might only be suitable when mixed with Scottish coal. In summary, the points he emphasised were:

1. That to be of value tests of coal must be on a large scale;

2. That success in using Irish coal would not only effect important economies, but also develop an Irish industry; and

3. That the assumption that the tests were worth making was based on the premise that Irish coal could be bought much cheaper than imported coal — an assumption that O'Brien noted could be false.

In conclusion, he drew the GS&WR's attention to the large use in the United States of very finely pulverised coal both for locomotive and stationery steam plant, adding that J.G. Robinson of the Great Central Railway had for some time been experimenting with a locomotive using pulverised fuel. Here, he added that he had discussed this question with Robinson the last time he had seen him and that he had been offered the opportunity of seeing the locomotive and riding with it at any time he cared to do so. H.E. O'Brien felt that the pulverised form might be one in which it would be possible to use Irish coal, its friability and large ash content not being detrimental in any way.

O'Brien's letter was forwarded to E.A. Watson, the Chief Mechanical Engineer, to prepare a reply informing the Ministry of the results of any experiments conducted by the GS&WR with Irish coal for general observations. The draft reply prepared by Watson was submitted and approved at the Traffic & Works Committee meeting held on 5th March. In it Watson advised that Irish coal had been tried on the GS&WR by practically all of the locomotive engineers since the line was opened. He noted that the most recent trials had been conducted in 1913 and 1914 with a train of six 30-ft coaches. In 1913 it was found impossible to maintain boiler pressure with the ordinary grate when using coal from the Wolfhill colliery in Co. Kildare however, when mixed in equal quantities with Welsh coal an improvement was achieved, but not sufficient to keep the boiler pressure high enough to maintain the necessary speed. In 1914 a Galloway-Hill patent grate had been fitted and this showed a distinct advantage over the previous trial, but not enough to make it possible to work average trains.

Watson went on to note that he had been in correspondence for a considerable time with an engineer in America and with J.G. Robinson with a view to ascertaining if Wolfhill coal could be used to advantage in pulverised form. He had also obtained a quotation for a coal pulverising plant but he did not consider that he could justify the very high cost of experimenting, especially as extensive experiments were being made on the Great Central, the results of which he was keeping in touch with. Watson also advised that he was sending a sample of Wolfhill coal to Robinson for a trial during the summer of 1920. Finally, he noted that the GS&WR had also tried coal from the Arigna colliery in Co. Roscommon and he was of the opinion that this coal would give satisfactory results if sufficient quantities could be supplied. It is not known whether Watson's reply sufficed to put the matter to bed, or whether the turbulent political scene in 1920s Ireland brought all such correspondence to an end, but there the official question of the use of Irish coal for locomotive purposes rested.

As a result of a meeting that took place between the management of the Bessbrook & Newry Tramway Company (B&NT) and Joseph Ingram, respecting rates and the need to repair the passenger rolling stock and replace the out-of-date electricity generating plant, H.E. O'Brien visited Bessbrook on 22nd March. He inspected the cars and the generating plant and presented his report on 3rd May. In it he advocated replacement of Motor Car No.1, which had been built in 1886 by the Ashbury Carriage & Iron Company of Manchester. However, he considered that the generating plant would work satisfactorily for a further two or three years. The B&NT was not happy with the report and wrote to the Ministry threatening to close down the tramway. As a result, it was agreed that the entire cost of replacing the generating plant would be met by the Government, but renewal of the cars would remain the responsibility of the B&NT.

In submitting written comments on the paper entitled *The Possibilities of employing Electric Motive Power on Irish Railways*, read by P.A. McGee before the Institution of Civil Engineers of Ireland on 3rd May 1920, H.E. O'Brien ventured to suggest that the author was seriously at fault in his financial calculations. He went on to point out that transmission lines and track equipment would cost at least £2,000 per mile each; locomotives £10,000 each; and that generating stations would cost £30 per kW and sub-stations £12 per kW. On that basis he calculated that the total cost of electrifying 3,000 miles would be £21 million at the prices of the day. Furthermore, he estimated savings from sale of redundant steam locomotives and credits for use of a proportion of the transmission system to supply lighting and industrial load would only yield about £5 million; resulting in a net capital expenditure of £16 million.

Turning to the question of operating costs, H.E. O'Brien noted that the only possible saving would be in coal consumption, as he considered that the National Union of Railwaymen would insist on retention of two men on each locomotive. Savings on coal would probably yield about 4.7% on capital cost, but he considered that a return of 15% was required to make the case for electrification of a mainline. To hammer home his point, he stated that the whole cost of locomotive, carriage and wagon renewal, repair and running costs in Ireland would have to be saved to make Irish railway electrification a commercial proposition! He concluded by stressing that the problem in justifying electrification was one of finding a line in which continuous use could be made of track and locomotives in order to get a return on those two main items of capital expenditure, and in arranging working of the stock to obtain such use.

We have already noted in Chapter Four that competition generated by electrification of the tram route between Dublin and Dalkey in 1896 had moved the DW&WR to consider the application of electric traction for their railway line between Amiens Street and Kingstown (now Dun-Laoghaire). In November 1896 the DW&WR had published a notice of its intention to seek powers to equip the line between Dublin and Kingstown as an electric railway, but nothing ever came of this. Neither did the DW&WR take up a proposition made in 1900 by Messrs Bennett & Ward-Thomas of Manchester, to draw up a scheme for electrifying a part of the company's system. We have also seen, in Chapter Five, that the question of electrification was still very much a live issue for the DW&WR in early 1904 when their Locomotive Superintendent, Richard Cronin, had visited the L&YR and other railways in England that were adopting electric traction. Following the introduction of an express tramway service between Dalkey and Dublin in early 1909, the Dublin & South Eastern Railway (D&SER) — as the DW&WR had become in 1907 — considered the possibility of electrification once again. Even though it was recommended that Charles H. Merz, who had designed and carried out the electrification of the NER lines, should be asked to look into electric working of D&SER suburban lines, the matter progressed no further, and it was to be another ten years before the question of electrification was again considered.

On 24th June 1920 H.E. O'Brien wrote from the Ministry to the D&SER seeking information for the Harcourt Street — Greystones and Amiens Street — Bray routes, which he needed "to make a rough approximate estimate of the cost of electrification".[98] George Wild, who had succeeded Richard Cronin as Locomotive Engineer of the D&SER in 1917, supplied the data requested. In his reply Wild drew attention to the most severe grades and permanent speed restrictions on each route, finally noting that all up trains on the Harcourt Street line had to make a mandatory stop at Ranelagh.[99] Eoghan then set about producing figures for electrifying the line from Amiens Street to Bray. Basing this on a half-hourly service for ten hours per day with two-coach sets (140 seats) and a twenty-minute service for three hours with four-coach sets (280 seats), he estimated that capital expenditure would be in the region of £220,000. This figure included eleven motor cars at £5,000 each and fourteen trailers at £2,500 each. [100]

On 30th June 1920 Sir Philip Dawson read his paper entitled *Electric Railway Contact Systems* to the I.E.E. In contributing to the discussion, H.E. O'Brien noted that from a railwayman's point of view both the overhead system and the contact rail system had disadvantages, but that the overhead system had the very serious disadvantage of the difficulty in finding places to erect the supporting structures. He was anxious to learn from Sir Philip what difficulties he may have encountered in that regard, and if he could provide information on the tonnage of structural steel that was used per mile and what percentage of the cost of electrification the overhead structure represented. In reply, the author stated that in the prevailing conditions of the cost of materials and labour he found it rather difficult to give a definite answer on the cost of structures, but he advised that experience gained

---

98 Letter from Col. H.E. O'Brien, Chief Mechanical Engineer, Ministry of Transport, Dublin, to George Wild, Locomotive Engineer, D&SER

99 The mandatory stop for all 'up' trains at Ranelagh was introduced as a result of the Harcourt Street accident on 14th February 1900, when an 'up' cattle special failed to stop and ran through the end wall of the station.

100 Estimated Cost of Electrification of D&SER, Amiens Street –Bray Section, June 1920

in the previous ten years had enabled a design to be prepared for construction of overhead equipment which was to a great extent independent of obstructions by junctions, points, crossings or anything of that kind.

Another matter that H.E. O'Brien became interested in whilst at the Ministry was the question of superheating five locomotives that the Belfast & County Down Railway (B&CDR) had on order from Beyer Peacock. This is hardly surprising as the L&YR had been at the forefront in applying this feature to improve locomotive efficiency. Although G.J. Churchward had been the first to fit a Schmidt superheater, when GWR 4-6-0 No.2901 *Lady Superior* was built new in May 1906, he subsequently adopted his own Swindon design with a low-degree of superheat. George Hughes equipped two of his 0-6-0 goods locomotives with Schmidt superheaters in November 1906 and from then on Horwich progressed towards being a supporter of high-degree superheat.

Eoghan O'Brien was keen to see this means of improving efficiency applied on one of the smaller Irish railways. The Great Northern Railway of Ireland, GNR(I), was the first Irish railway to experiment with superheating. Although their initial experience with 'Phoenix' smokebox superheaters was far from satisfactory, the application of, firstly Schmidt, and later Robinson superheaters from 1913 onwards gave far superior results. Their 'U' class 4-4-0 light passenger locomotives of 1915, with Robinson superheaters and 8-in. piston valves, were essentially a tender version of the saturated 'T' class 4-4-2 tank locomotives delivered two years earlier. Recognising the marked improvement in the performance of these 4-4-0s, the GNR(I) were quick to draw up a specification for a superheated version of their 4-4-2T, five of which were ordered from Beyer Peacock in 1919 along with fifteen superheated goods engines.[101]

For the purpose of comparison, O'Brien organised an exchange of locomotives between the GNR(I) and B&CDR and on the 19th July 1920 the GNR(I) sent 'U' class 4-4-0 No.198 to the B&CDR in exchange for their 4-4-2T No.17 for the day, but no record has survived of the outcome of the trials. The fact that the B&CDR got an exemption in respect of the four 4-4-2 tank locomotives on order probably indicates that superheating gave no distinct advantage on short suburban runs.[102] On the other hand, it was generally accepted that superheating was beneficial for longer heavier working and H.E. O'Brien was somewhat insistent on having the 0-6-0 goods locomotive so treated.[103] In the end it turned out that the addition of a superheater would have increased the weight of the locomotive to such an extent

---

101 The five 'T2' class tank locomotives, Beyer Peacock Order No.01881, Works Nos. 6035-39, were delivered to the GNR(I) as their Nos.1-5 in August 1921.

102 Beyer Peacock Order No.01930, Works Nos. 6073 & 6074 and Order No.02045, Works Nos. 6097 & 6098 delivered to the B&CDR in 1921 as their Nos.13, 18, 19 & 21 respectively.

103 Beyer Peacock Works Order Nos.01925 (engine) and 01926 (tender), Works No. 6072, delivered to the B&CDR as their No.4 in 1921.

that the axle loading would have become unacceptable for the B&CDR, and so the matter was dropped.

Belfast & County Down Railway 4-4-2T No.17, one of the locomotives involved in the exchange of 19th July 1920 - [Courtesy of the late Desmond Coakham]

*Great Northern Railway (Ireland) 4-4-0 No.199 (later named **Lough Derg**) in the black livery in which it entered service in 1915. This was delivered at the same time as No.198 of the same class, the other locomotive involved in the exchange of 19th July 1920. - [© IRRS collection]*

The locomotive exchange between the GNR(I) and the B&CDR was Eoghan O'Brien's last task at the Irish branch of the Ministry of Transport. On 22nd July the Ministry had signified their intention of releasing him back to the L&YR, but they requested that he should remain available for occasional duty, which it was anticipated would not unduly interfere with his responsibilities at Horwich.

Return to Horwich

Construction of 0-8-0 goods locomotives with large superheated boilers and Belpaire fireboxes had resumed at Horwich in November 1918, and nine had been completed when Eoghan O'Brien resumed duty at the Works in May 1919. New construction was dedicated to only that type of locomotive until January 1921, when the total number built in the post-war period had reached forty-five. His absence from Horwich whilst on loan to the Ministry of Transport from 1st January to 31st July 1920 has been recorded in Chapter Ten. With his return to Horwich on 1st August George Hughes had his top team back together once more. One of the first tasks that they set about was the rebuilding of the four-cylindered 4-6-0 express passenger locomotives that had been built in 1908-09. The original design had been produced when Zechariah Tetlow was Chief Locomotive Draughtsman and incorporated slide valves operated by Joy's valve gear, which was a standard design then in use on the L&YR, and the twenty locomotives were also built without superheaters. They were mediocre performers, proving to be poor steamers and heavy on coal consumption, and any superiority over their smaller predecessors, the Aspinall Atlantics, was imperceptible.

The big locomotives, which on account of their size were referred to as Dreadnoughts, had proved an embarrassment, but the intervention of the War had certainly prevented any major work being undertaken to improve their performance. With the brilliant J.R. Billington in charge of the locomotive drawing office and H.E. O'Brien back in the workshops, the time was opportune for Hughes to do something to resolve the problems. The design was completely re-worked and incorporated the latest thinking; new 16½in. x 26in. cylinders with 9in. diameter piston valves actuated by outside Walschaert's valve gear, a 28-element superheater, and a commodious side-window cab. Co-operating closely with Billington, O'Brien ensured that the task of rebuilding No.1522 was carried out to exacting standards. It emerged from Horwich in its rebuilt form at the beginning of November 1920 and was allocated to Southport shed. It soon showed a noticeable superiority over the original design, especially in free running at speed and reduced coal consumption. A total of fifteen of the original locomotives were rebuilt before Horwich turned its attention to constructing new locomotives to the same design from mid-1921 onwards. However, further problems were to emerge with the re-designed 4-6-0s, and these are related later in this chapter.

On 8th October 1920 H.E. O'Brien delivered his paper entitled *The Management of a Locomotive Repair Shop* to the Manchester Centre of the I.Loco.E. Twelve days later he read it again at the London headquarters, and on 26th November he presented it in Glasgow as the first paper to be read before the newly formed Scottish Centre. The paper described the engineering principles by which Horwich had been guided for many years, and in it he pointed out that modern high-speed machine tools had to be worked to full capacity if an adequate return was to be obtained on the investment made in the same. He added that the intelligent manufacture of parts for stock was also important, and that knowledge of the minimum and maximum weekly or monthly demands for each item was required, which had to be supported by accurate stock control and issuing procedures. Other salient points that he felt needed the closest possible attention were:

(1) Reduction of unnecessary work to a minimum by:

   (a) Inspection of all parts for flaws and wear after they had been stripped and cleaned;

   (b) Elimination of all unnecessary machining operations;

   (c) Elimination of labour-intensive fitting operations by the use of limit gauges, jigs, and fitting by grinding instead of scraping, filing or chipping.

(2) Reduction of unnecessary transportation of material by:

   (a) Arranging for all white metalling and coppersmiths' work to be carried out in closest proximity to the Erecting Shop;

   (b) Grouping machines and work benches dealing with the respective machined items in the same vicinity;

   (c) Retention in the Erecting Shop of all material that was proved by inspection not to require repair.

(3) Effecting the closest economy in the use of material by:

   (a) Specialised inspection of items for wear;

   (b) Noting parts where wear can be reduced by use of hardened and ground surfaces;

   (c) Extensive use of graduated sizes of pins and holes and by the use of renewable hardened and ground bushes;

   (d) Regular inspection of all scrap material with a view to utilising for manufacture of new articles of a smaller size;

   (e) Close analysis of the causes leading to renewal of any parts for which appreciable quantities are used annually;

   (f) Elimination of the more expensive metals wherever possible;

   (g) The judicious use of welding and patching.

H.E. O'Brien went on to note that 80% of locomotives stopped for repair work were taken out of service on account of boiler condition, and he emphasised the importance of statistical information in monitoring trends and recording operational and repair information. He felt that this was the only way to maintain control, but he cautioned that statistics should not be allowed to 'run riot' and create a need for a large number of inspectors and clerks to produce information that would serve no useful purpose. He also emphasised the few cardinal points that had always been kept prominently before the management at Horwich, viz.:

(1) The necessity for keeping the boiler stock young — the L&YR operated on a boiler life of 16 to 17 years;

(2) The desirability of reducing firebox patching to an absolute minimum;

(3) Making the greatest possible use of machinery in order to economise labour;

(4) Adoption of piecework, or payment by results, together with an endeavour to secure the reasonable and intelligent co-operation of labour in obtaining maximum output;

(5) The infusion of a spirit of broad-minded economy and esprit-de-corps into the supervisory staff; and

(6) The most vigorous adherence to standards.

A lengthy appendix to the paper described in detail the system of locomotive repairs in use at Horwich, and a second appendix dealt with the application of limit gauges and progressive sizes in the Machine Shop.

For the discussion in London contributors included Basil Kingsford Field (Works Manager, Brighton, LB&SCR). At Manchester, on 5th November, many of his own colleagues, including F.W. Attock, J.R. Billington and J.H. Haigh, participated along with J. Parry (Great Central Railway, Gorton) and, in Glasgow, Walter Chalmers (Locomotive Superintendent of the North British Railway) and Irvine Kempt (Caledonian Railway, St. Rollox) contributed. The significance of this seminal paper has, perhaps, been overshadowed by the subsequent emphasis on H.E. O'Brien's contribution to mainline electrification. However, he was, first of all, a trained mechanical engineer, and the importance of his contribution was recognised by his peers. At the 11th Annual General Meeting of the I.Loco. E., which was held at Caxton Hall on 23rd March 1922; the Secretary announced the award of 1st prize to H.E. O'Brien for his paper *The Management of a Locomotive Repair Shop*. As late as 1975 Roland Bond was to recall in his autobiography that the principles outlined were "as valid today as when they were first enunciated", and he reiterated what Col. O'Brien had written, viz.:-

> "It should be noted at the outset that the main objective of the management of a railway locomotive workshop is essentially different from that of a commercial manufacturing works; the engineering management of a commercial engineering works desire to see a constant expansion of their shops, while in the case of the railway management their desire should be to see a constant shrinkage of the shops brought about by: (i) Improved methods of manufacture; (ii) Improved organisation; and (iii) Rectification of errors of design and material with the object of reducing renewals and repairs to a minimum.
>
> "It is possible to effect this because the capital expansion of the locomotive stock

is very slow on English railways, and therefore the capacity of the Works should more than keep pace with the demands if the management is progressive, in spite of increased weight and power of the more modern stock. The number of locomotives on the L&YR had only increased from 1,326 to 1,645 in the previous twenty years. The aim of every railway management should therefore be to effect improvement in the cost of repairs at a greater rate than the capital expansion, so that far from requiring additional shop room, increased floor space will steadily become available in the existing shops while the number of employees remains constant or even decreases."[104]

To understand the next phase in Eoghan's career, we must return for a moment to 1919. After the War, the British railway companies found that they were physically exhausted by the stresses, strains and losses of the previous four years. They had also suffered from inadequate financial compensation for their burden of extra work while under Government control during the war years. With the powerful new Ministry of Transport formed under Sir Eric Geddes in existence, a White Paper was circulated proposing that 123 railways should be reorganised into seven groups. The companies agreed in principle but disliked the proposed Parliamentary groupings and suggested a new grouping into five companies, four regional groups plus a grouping of local railways in the London area. Parliament subsequently adopted that approach with the exception of the London local grouping. Under the Railways Act 1921, which became law on 19th August, a total of 130 railway companies were grouped into four new entities, the GWR, SR, LMS and L&NER. They were handed back from Government control with a £60 million payment to their owners as full and final settlement of all Government liability incurred between 5th August 1914 and 31st December 1918.

Merger arrangements had to be completed on or before 1st January 1923, but any pair of railway companies designated as being within the same group could voluntarily amalgamate earlier, subject to the approval of the Amalgamation Tribunal established under the Act. Pending the grouping, the L&NWR and the L&YR decided to take advantage of this arrangement and agreed to amalgamate under the title of the former company with effect from 1st January 1922. That merger had been foreshadowed with the appointment of Sir Arthur Watson, General Manager of the L&YR, to be General Manager of both companies from 1st January 1921. Upon the merger coming into effect, George Hughes was appointed Chief Mechanical & Electrical Engineer of the enlarged L&NWR and, with the former electrical department of the old L&NWR also under his control, H.E. O'Brien was appointed Electrical Engineer at a salary of £2,500 per annum, the amount taking into account an allowance for pupils' fees.

Contributing to the discussion on the paper *Rotary Converters with special reference to Railway Electrification*, read by F.P. Whitaker before the I.E.E. on 16th February

---

104 Bond, R.C., *A Lifetime with Locomotives*, Goose & Son, Cambridge, 1975

1922, H.E. O'Brien related the L&YR experience with their 1,350V, 25Hz rotary converters that had been installed for the Manchester-Bury electrification in 1915. Noting that a great deal of progress had been made since that date, he stated that considerable problems experienced with the machines in the early days had been overcome when the compound winding was taken off. However, he was of the view that his fortunate experience with the rotary converters was probably due to the batteries which the L&YR had installed at their sub-stations, and he welcomed any apparatus designed to prevent flash-overs in the machine as described in the paper.

Eoghan had begun to build up his team at Horwich when appointed Electrical Engineer in January 1922, and almost immediately he started to investigate areas where electric traction could deliver a distinct advantage for the 'greater L&NWR'. The L&NWR section of the West Coast mainline from Crewe to Carlisle included the arduous climb to Shap summit, 916ft. above sea level. Northbound trains had to face a continuous climb of 5¾miles, the last four consecutive miles of which were at 1 in 75. The Southbound journey was even harder work for locomotives with a continuous climb of over 10 miles including 7 consecutive miles at 1 in 125. This presented an obvious target for electric traction and H.E. O'Brien obtained authority to proceed with trials to assist in preparing a case for electrification of the Crewe-Carlisle route. Matters had advanced quickly enough for E.S. Cox (at that time a draughtsman in the Horwich drawing office) to be preparing outline diagrams of the proposed electric locomotives for the service by the time the LMS came into being on 1st January 1923.

Appointments to the senior positions in the LMS, which were to come into effect on 1st January 1923, had been announced on 16th December 1922, and George Hughes became the Chief Mechanical & Electrical Engineer of what was the largest of the 'Big Four' railway companies. H.E. O'Brien was appointed Electrical Engineer of the LMS on the same salary and conditions that he had enjoyed on the 'greater L&NWR'. In his role in the enlarged organisation he took as much interest in staff recreational activities as he did with engineering matters, and was among the officers who were present at the summer meeting of the LMS (London) Athletic Club, which was held at Wembley on the first Saturday in June 1923.

Work on the Crewe-Carlisle project continued and, in order to obtain first hand data for the specification of the electric locomotives, two special dynamometer car tests were made over the proposed route. The first test run was made on 24th June 1923, when one of the rebuilt L&YR 4-cylinder 4-6-0s, No.1511, together with one of the new locomotives built to the same design, No.1658, double-headed a 500-ton train from Crewe to Carlisle. The run was marred by a somewhat indifferent performance by both locomotives. The second trial run was undertaken on 18th November when two of the new locomotives of the same class, Nos.1669 and

1670, hauled a 640-ton train in both directions over the Preston-Carlisle part of the route. This was a much better performance as confirmed by the draw-bar pull and horsepower charts, which were subsequently included in H.E. O'Brien's paper to the I.E.E. entitled *The Future of Mainline Electrification on British Railways*.

The first reading of the aforementioned paper was made before the Mersey & North Wales Centre of the I.E.E. on 17th March 1924 at Liverpool. On the following day he read it before the North-Western Centre at Manchester, and the North-Eastern Centre heard it on 24th March. Col. O'Brien presented the paper at Savoy Place, London, on 27th March and finally before the South Midlands Centre on 2nd April. It was a masterpiece, and established the case for mainline electrification to an extent that had not been achieved up to that time. The contributors to the discussions held at the various centres read like a veritable 'who's who' of railway electrical engineers, but perhaps one of the most important, albeit the most brief, contribution was that made by J.R. Billington who stated, "I can vouch for the accuracy of the dynamometer car figures which the author has so ingeniously used, together with other railway statistics, to prove his contentions".

The O'Brien trial double-headed by Hughes' 4-6-0s storms over Shap - HOR-F-3541 © NRM, York]

In the discussion of the paper at the I.E.E. headquarters, Savoy Place, Roger T. Smith, I.E.E. President, noted that the comparison between steam and electric locomotives for various traffic densities was based on assumptions that were quite agreed among electrical engineers. He went on to point out that the author had not mentioned the great traffic advantage offered by the speed characteristic of electric motors as compared with that of the steam engine, which had been well

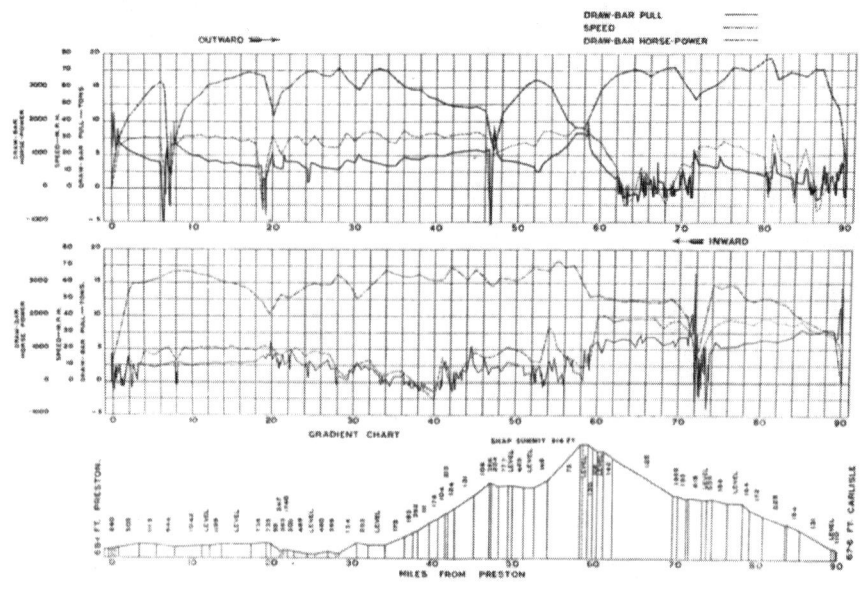

Power output recorded during O'Brien's double-headed trial run between Crewe and Carlisle with a 500-ton train. - [Reproduced by permission of the Institution of Engineering and Technology]

demonstrated on the Italian State Railways. In view of the advantages of main-line electrification he noted that the French intended to electrify 5,600 track miles; the Italians 2,800 track miles; and the Swiss 1,000 miles, and he queried why there was comparatively no main-line electrification in Britain. In answering his rhetorical question R.T. Smith proffered two reasons: (1) in the greater part of the countries in question coal was an import, and (2) traffic officers in Britain were still not convinced of the merits of electrification.

Others who contributed to the discussion included F.W. Carter, Francis Lydall (Merz & McLellan), and James Dalziel (LMS and formerly of the MR). The opening statement of Dalziel's contribution probably best summed up the admiration with which the paper was received:

> "That the author's investigations show a more favourable case for electrification from the financial point of view than some of us have hitherto been able to attain is no doubt due in part to the closer estimates that he has been able to take out in the light of the information at his disposal, and in part also perhaps to his being able to put a valuation on advantages which have hitherto had to pass by. I do not think, however, that he can be accused in any quarter of using figures that he cannot fully substantiate or that railway or electrical engineers generally would not be prepared to accept. That the paper is centred on finance and takes technics largely for granted not only is wise, but marks a stage in the literature of electrification."[105]

Contrary to previously published accounts, there is no evidence to indicate that it was his I.E.E. paper that was to cause discomfort among the LMS hierarchy. It

---
105 J. Dalziel in discussion of The Future of Mainline Electrification on British Railways, I.E.E., 1924

CHAPTER TWELVE Return to Horwich

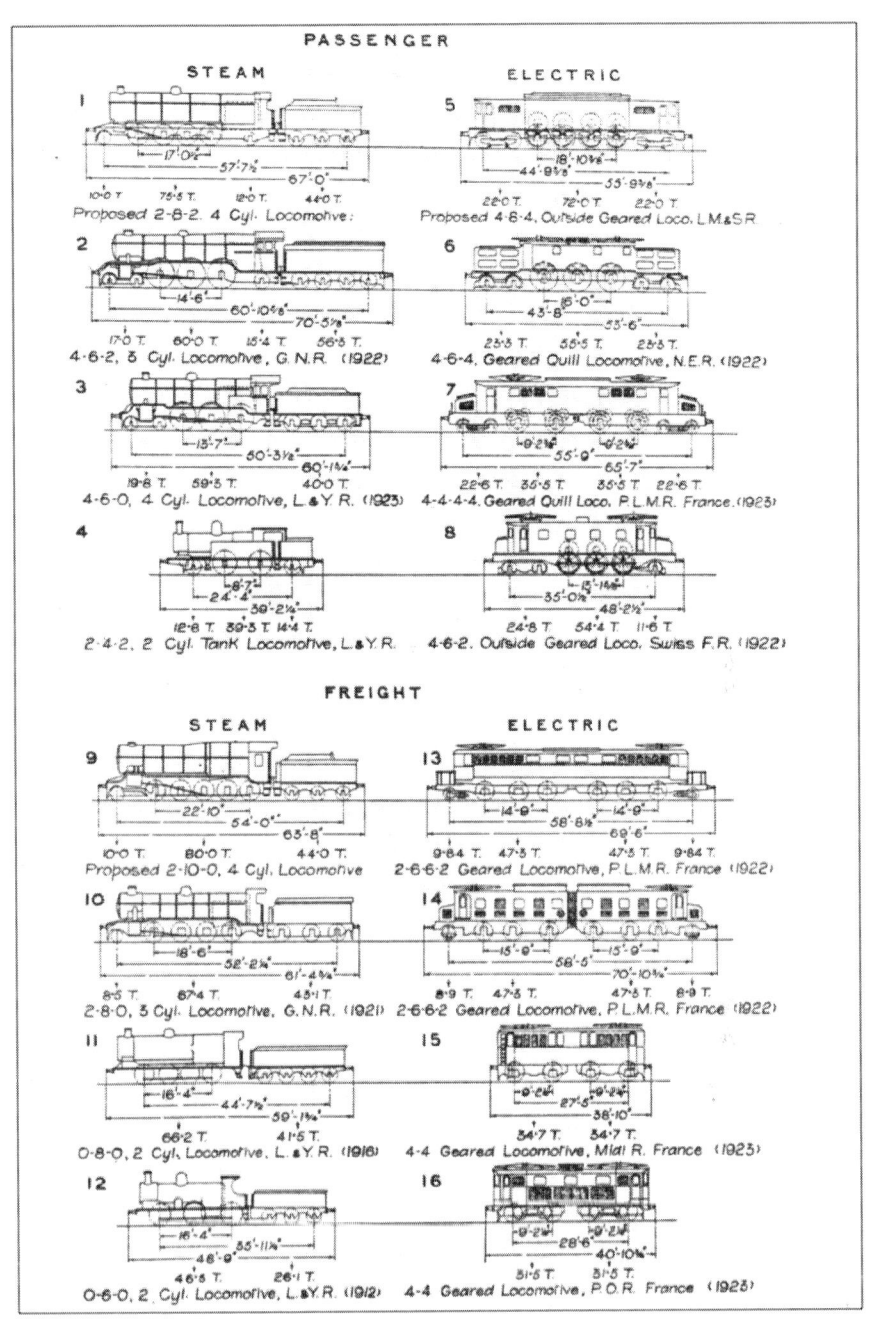

Comparative diagrams of steam and electric locomotives as illustrated in Col. O'Brien's paper The Future of Main Line Electrification on British Railways presented at Institution of Electrical Engineers - [Reproduced by permission of the Institution of Engineering and Technology]

had been prepared and presented with Hughes' blessing and it is most unlikely that the Chief Mechanical Engineer would not have given his approval without having first cleared the matter with his superiors. In fact, the time that elapsed between the presentation of the paper in March-April 1924 and the subsequent hiatus at the end of the year suggests that it was a later event that was to be the catalyst of disagreement of certain elements on the LMS Board with Eoghan's views. However, before considering the particular circumstances that surrounded his departure from the LMS, we must return to 1923 to review other activities in which he had an involvement.

H.E. O'Brien made a contribution to the discussion on *Twelve Years' Operations of Electric Traction on the London Brighton & South Coast Railway*, a paper read by H.W.H. Richards to the I.C.E. on 9th January 1923. He was also involved in preparing for the supply of electricity from the power station at Formby to the Urban District Council. On 31st May 1923 arrangements were confirmed at the rate of 1¾d per unit between midnight and 5.0 a.m. on weekdays and midnight and 9.0 a.m. on Sundays, Christmas Day and Good Friday, and at 4.0d per unit during all other hours. On the same date he attended the Rolling Stock Committee of the LMS Board for the first of four occasions that he was to do so, the other occasions being on 28th June, 29th November and 20th December 1923. At the meeting held on 29th November the locomotive building programme for the winter of 1923-24 was discussed, and most interestingly it included a quantity of one hundred 2-8-0 goods locomotives at a cost of £5,000 each. In regard to those locomotives, Hughes explained that in consultation with the Chief General Superintendent, J.H. Follows, he was endeavouring to design a particular type of locomotive to meet certain requirements, but he was not sure at that stage what particular type would be required to operate over all sections of the LMS.

At the Committee meeting held on 20th December 1923, O'Brien presented alternative estimates for the electrification of the Manchester-Oldham branch. He calculated that if some of the current required was obtained from Oldham Corporation and Manchester Corporation the capital cost would be £299,650, but if the whole of the current were to be generated at Clifton Junction an extension to the power station would be required and that the overall capital cost would rise to £467,350. Approval was given to obtain a supply of electricity from outside sources and to install ¼mile of 3rd rail under contract for experimental purposes, but it appears that nothing more was ever heard of the scheme.

Also in 1923, Monsieur André Bachellery, Chief Engineer, Chemins de fer du Midi, read a paper on *Electrification of the French Midi Railway* to a Joint Meeting of the I.E.E. and the British Section of the Société des Ingénieurs Civils de France. Contributing to the ensuing discussion in London on 22nd November 1923, H.E. O'Brien commented that what had been done in France in regard to the standardisation of voltage and appliances in connection with electric traction should be an example for Britain. He noted that both the French Government and the French electrical industry had appeared to have grasped the problem boldly, and that the Midi Railway, which had spent a very large amount on single-phase electrification, was, for the good of the country, ready to abandon that system and adopt the national standards. In regard to electrification of mainlines in Britain, he regretted that railway management had up to that point only dimly appreciated the potential. Noting that the fundamental question was the cost at which electric current could be supplied to the locomotive, he pointed out that the author had the distinct advantage of hydro-electric power at his disposal, and that his cost of energy would tend to become cheaper as usage increased, whereas in the U.K. it would not decrease at the same rate.

In March 1924 the Rolling Stock Committee of the LMS was split into two separate committees, the Locomotive & Electrical Committee and the Carriage & Wagon Committee.[106] At the first meeting of the Locomotive Committee held on 30th April, Hughes reported that it had not been possible to arrive at a design of 2-8-0 locomotive that could be utilised on all sections of the line, and that in view of the urgency of the J.H. Follow's requirements he had agreed with him that 100 superheated 0-6-0 'Class 4' goods locomotives should be substituted instead. By the time of the meeting of 23rd July it had been agreed to include one hundred 2-6-0 mixed traffic locomotives of a new type in the locomotive building programme for the following year (1924-25).

H.E. O'Brien was in attendance for the Locomotive & Electrical Committee meeting held on 25th June 1924, when the subject of flash-overs in the Manchester-Bury rotary converters came to the surface again. Reporting on the matter, he advised that not only were the flash-overs a cause of expense in the maintenance of the brush gear, but that they also posed a danger to the attendants at the sub-stations. He stated that five of the eight single-commutation 1,200V machines had been altered by the fitting of intermediate pole pieces and a damping ring, which had resulted in a 50% reduction in the number of interruptions to supply. He went on to recommend that the three remaining machines be altered at an estimated cost of £1,500, which was approved.

Sir Arthur Watson resigned in 1924 due to ill health and was succeeded as General Manager of the LMS by H.G. Burgess, whom we have met before in Chapter Ten. In the same year Sir Guy Granet succeeded Lord Lawrence as Chairman. Granet, a wily old lawyer and formerly Chairman of the MR who, according to Hamilton Ellis, "would have out-manoeuvred Machiavelli himself"[107] had virtually dictated the terms of the amalgamation of the companies that formed the LMS. As one of the two joint deputy chairmen he had subsequently dominated the proceedings in the early years of the LMS and had a freer hand to do so following his appointment as Chairman. The result was that the newly formed company adopted the MR

---

106 LMS Board Minute No.594, dated 27th March 1924
107 *The Midland Railway*, Ian Allan, 1953

precepts of management.

In 1906 Granet had completely overthrown the MR management organisation, which had followed the traditional form that prevailed on all large railways in Great Britain up to that time. He appears to have been one of those people bent on reducing the status and power of engineers. Once Granet had become General Manager of the MR, Richard Mountford Deeley, the Locomotive Superintendent, had found that his plans to introduce appropriate modern motive power for heavy freight and crack expresses trains were getting nowhere. Perhaps somewhat more significant to later events was the fact that Deeley, who was also an electrical engineer of distinction, had pioneered single-phase electrification at 6,600V, 25Hz with overhead contact on the routes from Lancaster to Heysham and Morecambe, which were brought into service in 1908. Not long after, differences appear to have arisen with Granet, and one morning in 1909 Deeley stormed out of Derby Works never to return again, it being generally accepted that 'the dead hand of Derby' in locomotive matters can be traced back to that time.

On 25th September 1924, H.E. O'Brien read a paper entitled *Main Line Electrification* to the I.Loco.E. in London. This was not a re-run of the earlier masterpiece presented to the I.E.E., but a paper on the distinct advantages of the electric locomotive compared to the steam locomotive. Bearing in mind that he was probably preaching to the unconvertible — representatives of the steam locomotive building industry including Col. E. Kitson-Clark (Kitson); W. Cyril Williams (Beyer Peacock) and R.P.C. Sanderson (Baldwin Locomotive Works), to mention just three — it is hardly surprising that Eoghan met with hostility, albeit politely expressed in the ensuing discussion. But, perhaps the more significant feature of that evening's proceedings was that James Edward Anderson, the LMS Superintendent of Motive Power, who at the time was President of the I.Loco.E, took the chair.

Sir Guy Granet, Chairman, LMS (1924-27) [Reproduced from the LMS Centenary issue of Railway Gazette September 16, 1938, by permission of the Editor]

We have already noted that the Midland had gained supremacy at Board level of the LMS, and under the new management structure motive power affairs were divided between Hughes, the Chief Mechanical Engineer, who was responsible for design, and overhaul; and Anderson who was in charge of all locomotive running matters including maintenance activities in the running sheds. Under this arrangement Hughes had lost one of his right hand men, F.W. Attock, who from then on reported to Anderson as Divisional Motive Power Superintendent, Central Division, Manchester. Anderson, who had stood in as Chief Mechanical Engineer on the Midland Railway during

Fowler's war service with the Ministry of Munitions, may have considered that he held the position that really mattered when it came to motive power decisions — he certainly acted as if he did! He also appears to have had the ear of Sir Guy Granet, his former General Manager from MR days. One way or another, it was following the presentation of his paper to the I.Loco.E. on *Main Line Electrification* that H.E. O'Brien found himself called to the Board Room to account for his opinions. As there is no formal record in the Board Minutes regarding what took place, it must be assumed that it was a private meeting between the Chairman and H.E. O'Brien, with possibly one or two other Board members present.

The rivalry that had existed between the MR and the L&NWR for west coast traffic prior to the grouping may still have been foremost in the minds of certain Board members who were not whole-hearted supporters of the grouping. The GWR was, after all, hardly affected by the grouping in so far as its main routes were concerned; the three constituents of the SR continued to operate in their own territories; and there was really no alternative in the L&NER group to the GNR/NER/NBR east coast route to Scotland. However, Anglo-Scottish traffic from London via Carlisle could go by either L&NWR or MR routes. Despite external appearances of a unified company, the 'Midland' thinking inside the LMS was distinctly parochial. It was bad enough having to contend with the arduous Settle-Carlisle section, but the idea of Hughes' proposed locomotives providing an effective through service on the L&NWR route were an anathema to Midland minds. To add to that there was an electric traction proposal that would make Shap seem like child's play, and it didn't even require two men on the footplate of an electric locomotive! This may have been just too much for certain Midland minds; Horwich had to be stopped before the Euston-Carlisle route took any greater prominence than it already possessed within the new organisation.

As far as the Midland men were concerned, any new motive power acquired for the LMS had to be capable of operating within the restrictions of the MR infrastructure. The obvious solution was to continue with the production of standard MR locomotive designs; 4-4-0 passenger locomotives (Compounds and Class '2P') and the Class '4F' 0-6-0 goods locomotives. It was very suitable to Midland interests if the locomotive policy that had been pursued since Deeley's departure fifteen years earlier remained unchanged. The work on effective modern motive power solutions that Hughes was undertaking with his team at Horwich would not benefit the Midland route, and so it appears that behind-the-scenes moves were made to undermine his efforts. H.E. O'Brien's paper may probably have unwittingly provided the anti-Hughes camp with their first real opportunity to attack the Horwich team.

Whatever transpired at Euston it cannot have been too friendly. Eoghan O'Brien was accused of interfering in policy (undoubtedly Anderson's Midland-orientated small locomotive policy) and was told that he could forget ideas of mainline electrification. Anecdotal evidence suggests that he resigned in disgust, but there was probably no option. However, he did secure a good package; half his retiring salary for the rest of his life and, as we shall see, he certainly lived long enough to make that count! The deal also included a Gold Pass for first class travel in Britain and Western Europe for the rest of his life and a Silver Pass with similar conditions for his wife. The retirement deal was, however, subject to one condition — that he should not engage in any activity considered to be detrimental to the interests of the LMS.

Eoghan's retirement from the LMS on 31st January 1925 was noted at the Locomotive & Electrical Committee meeting on 25th February under *Re-organisation of the Electrical Department*. In his place Frederick Augustus Cortez-Leigh was appointed Electrical Engineer with James Dalziel as his assistant. However, a more far-reaching decision made at the meeting was that of relieving George Hughes of responsibility for electrical matters, ostensibly "in view of the magnitude of locomotive work". It was obvious that H.E. O'Brien's departure did not spell the end of the push against Horwich and the Derby clique must have felt that everything was really going their way when less than a month later J.R Billington, Hughes' Chief Locomotive Draughtsman, died at the young age of 49 on 22nd March 1925. Hughes had been deprived of responsibility and had lost two of his most senior and best men and a month later he had had enough. In a letter dated 21st April 1925 he gave notice of his intention of resigning his position as Chief Mechanical Engineer with effect from 15th November, on which date he would have completed 37 years membership of the superannuation fund and 43 years service.

It has been said that Hughes had a blind spot when it came to boiler pressure, but he was convinced that the longer boiler life that could be obtained using low pressure far outweighed the benefits that could be achieved by increasing the working pressure. In this regard he was no different to Gresley who had originally adopted the same boiler pressure of 180 lb/in^2 for his Pacifics. Billington's re-design of the Dreadnoughts had retained the 180 lb/in^2 boiler, but with a 28-element superheater and, although they incorporated Walschaert's valve gear and long-travel piston valves, they had a short lap. Trials over Shap against L&NWR 'Prince of Wales' and 'Claughton' 4-6-0s appeared initially to favour the improved Horwich product and thirty-five more, equipped with larger tenders, were ordered for West Coast services. However, their increase in coal consumption with mileage after repairs, and erratic steaming rendered them less suitable than the Claughton's for sustained effort. They also suffered from steam leakage past the Hughes patent piston valves, and loss of smokebox vacuum. The first problem was overcome by using re-designed valve heads with narrow rings, but the departure of Hughes brought a halt to any further improvements.

CHAPTER TWELVE Return to Horwich

What might have been – Robin Barnes' artistic impression of O'Brien's proposed 2-Do-2 electric locomotive heading south out of Lancaster with the L&NWR Royal Train set of 1903 -
[Courtesy of the artist]

Hughes 4-cylinder superheated passenger locomotive in LMS livery -
[Courtesy Railway Magazine]

Hughes' resignation sent ripples through the organisation and a sense of the Directors' appreciation of his services can be obtained from the minute recording the Chairman's statement, which read

> "Mr Hughes was appointed to his responsible position at the formation of the group and had, in setting up his organisation, and in carrying on the work of the department, to contend with extraordinary difficulties. The Chairman felt that it was the desire of

185

the Board that he should congratulate Mr Hughes on the success which had attended his efforts and express to him the appreciation of the Directors for the very valuable services that he had rendered to the company in very difficult circumstances, and the hope that he may be spared to enjoy his well-earned rest."[108]

It took until 29th July for a recommendation on a replacement to emerge. In the event, the appointment of Sir Henry Fowler was not as straightforward as we have always been led to believe. There seems to have been some reservation in giving him the job and, although he received a rise in salary of £1,000 per annum, his appointment was made on the strict understanding that "if within three years of the date thereof it should be decided by the company to carry out any scheme of reorganisation of the Chief Mechanical Engineer's department involving either his retirement or transfer to another position, any claim that he may have for compensation under the Railways Act, 1921, shall be settled on the basis of his present salary of £3,500."[109]

Although Fowler was a genial character, he had not really demonstrated any flair for innovation in the development of motive power on the MR, and his time on the LMS did not prove very much different. The fact that the LMS hierarchy did not select a dynamic replacement for Hughes in 1925 probably explains more than anything else why the same sort of efforts that Gresley made to improve his 'A1' Pacific locomotives were not applied in overcoming the residual problems with the Dreadnought 4-6-0s. The end of the Hughes era was marked with a farewell function held at Horwich on 11th November 1925, G.N. Shawcross making a presentation on behalf of the staff. The days of the Horwich triumvirate of Hughes, O'Brien and Billington were over, the Chief Mechanical Engineer following his long-serving right-hand man into retirement far away from Horwich; George Hughes on the east coast at Cromer and Eoghan O'Brien back at *Mount Eagle* in Killiney.

---

108 LMS Board Minute No,1289, dated 30th July 1925
109 Minutes of LMS Locomotive & Electrical Committee, dated 29th July 1925

Retirement at Killiney

It may be asked why H.E. O'Brien did not find a position with one of the other major British railway companies, particularly the L&NER where his first cousin, Murrough Wilson, was on the Board. The answer lay, firstly, in the terms of his settlement with the LMS which precluded such a move and, secondly, with the L&NER itself. Sir Vincent Raven acted in a consultancy role following the formation of the L&NER, but when that arrangement expired Gresley set about getting his own electrical expert and 'poached' Henry Walter Huntingford Richards from the Southern Railway, who was appointed as his Electrical Engineer from 1st July 1924. And so it was that Eoghan O'Brien returned to Killiney to enjoy retirement and his pastimes of mountain climbing, hill walking, salmon fishing, and bee-keeping. He was a member of three Alpine clubs (the British, French and Swiss) and, with the exception of the War years, had visited Switzerland annually since his student days.

Free from work commitments his enthusiasm for motoring resulted in the purchase of a 14hp Sunbeam in England for the purpose of a long motoring holiday with Frances in Algiers, Tunisia and Sicily during the summer of 1925. They crossed to Boulogne and drove down through France to Marseilles where they boarded a ship for the 26-hour crossing to Algiers. The return to Europe was made from Tunis to Palermo, and the couple spent nineteen days in Sicily before proceeding up through Italy to Switzerland. Professionally Eoghan was not idle for he also quietly pursued some interesting and amusing consultancy assignments until the outbreak of war in 1939; but, as we shall see, a good deal of this work was in 'behind the scenes' activities. He travelled all over Europe reporting on progress in electric traction, and was the author of *The economic aspects of Railway Electrification with special consideration of those which cannot be expressed numerically*, a paper that he read at the 11th session of the International Railway Congress Association held at Madrid in May 1930.

Although H.E. O'Brien was one of the railway representatives who had sat before the Electrification of Railways Advisory Committee in 1920, his departure from the LMS in January 1925 meant that he was not in a position to be consulted by the Railway Electrification Committee established by the Government on 11th November 1927 under the chairmanship of Col. Sir John W. Pringle. However, it appears that the situation may have changed somewhat by September 1929 when the Committee on Main Line Railway Electrification was appointed under the chairmanship of Lord Weir of Eastwood. Most importantly, Sir Ralph Wedgwood, one of Eoghan's close acquaintances from his years of war service, was one of the members of that three-man committee. Equally significant was the list of witnesses, which included many of his old colleagues and acquaintances including Sir John

Aspinall, Sir Henry Fowler, Nigel Gresley, and Charles H. Merz his close friend and relation by marriage.

In addition to his role as a witness, Merz was to play a significant part in the findings of the Committee as it was decided to arrange for Merz & McLellan to conduct two professional investigations into electrification of sections of British railway systems. The first scheme consisted of practically all that portion of the L&NER that was formerly operated as the Great Northern Railway, and the second scheme covered the Crewe to Carlisle portion of the LMS together with the Weaver Junction to Liverpool section and 33 miles of branches and loops. Not only was it a strange co-incidence that the Committee should consider the very same section of the LMS that had been proposed for electrification by H.E. O'Brien, but the Merz & McLellan arrangements bear a distinct similarity to his proposals.

Of significance in considering his unofficial involvement is the different tone that Merz & McLellan adopted in acknowledging the assistance rendered by the two railways. In the case of the L&NER they stated:

> "In carrying out this investigation we have kept throughout in close touch with the officers of the Railway, who, at the expense of much hard work, have prepared all the necessary statistics of traffic, estimates of alterations to ways and works, steam working costs and valuation of assets. They have also drawn up revised time tables for electric working, have calculated for us on the basis of this revision two important items of the electric working costs, and have advised us as to the number of electric locomotives of different classes necessary for dealing with the traffic. We wish to express our appreciation of the way in which this work has been done."

but in the case of the LMS, the wording was significantly different, viz.:

> "In conclusion, we would like to express our appreciation of the assistance which has been given to us by the officers of the Railway who have spent a great deal of time in studying the problem and in preparing the information we required for our calculations."

Reading between the lines, there is no doubt that the LMS still had a negative view on electrification, and under such circumstances it is not difficult to imagine a purely 'off the record' series of communications between Charles Merz and Eoghan O'Brien. Merz would have been well aware of the 1924 paper to the I.E.E. on the future of mainline electrification, and would have found it a most useful basis for the LMS investigation. The most striking similarity between the Merz & McLellan report and H.E. O'Brien's proposals is in the types of locomotives recommended for the Crewe – Carlisle trains. In both cases a 2-D-2 (4-8-4) electric locomotive is proposed; a 116 ton, 3,000 hp unit by O'Brien and a 100-115 ton, 2,400 hp unit by Merz & McLellan. Both papers proposed Bo-Bo locomotives for ordinary freight (1,800 hp in the case of Merz & McLellan) and a small number of lower horsepower Bo-Bo units for shunting and local freight. The only difference was in the proposals for fast or heavy freight trains, where O'Brien intended to use the 2-D-2 locomotive as standard, but Merz & McLellan opted for a fourth type of locomotive; a 2,100 hp Co-Co design of 108 tons.

There is no doubt that the inclusion of branch lines, particularly the re-equipping of the Lancaster – Morecambe – Heysham branch to convert it from single phase AC traction to 1,500-V DC traction, and the introduction of a fourth type of locomotive, altered the economics presented in H.E. O'Brien's paper. Despite the effect of these factors the Merz & McLellan report still demonstrated a difference in comparative working expenses of £127,766 per annum in favour of the electrification scheme for the LMS; equivalent to a 2.5% return on the net capital expenditure of £5,123,370. On the other hand, the L&NER scheme showed a 7.2% return.

We shall probably never know to what extent Eoghan was involved behind the scenes. However, in commenting on a paper by S.B. Warder on *Electric Traction prospects for British Railways*, read before the I.Loco.E. on 13th December 1950, Eoghan stated that in 1924 he had read a paper on this subject; of which, he believed, extensive use was made in compiling the Weir report of 1931.[110] Neither shall we know what difference a similar approach to that taken by the L&NER would have made to the reception of the LMS scheme, but even by 1931 the LMS authorities were unlikely to budge from their small steam engine policy no matter what case was put forward in favour of improving railway traction.

In Chapter Ten we noted the various attempts that had been made to introduce electric traction on the Dublin – Bray line and Eoghan's personal involvement during 1920 with proposals for electrification of this section of his 'home railway'. The D&SER Board again discussed electrification of suburban lines in July 1922, the General Manager being requested to resurrect the report obtained from the Ministry of Transport in 1920.[111] By this stage the Civil War, which had broken out in June 1922, was beginning to disrupt D&SER operations and nothing more happened in regard to electrification before the D&SER became part of the Great Southern Railways (GSR) on 1st January 1925. A few months later the Shannon Electricity Bill passed through its final stages in the Dáil[112] and Senate and on the 13th August 1925 the Free State government entered into a contract with Siemens-Schuckert for construction of a hydro-electric generating station on the River Shannon at Ardnacrusha.

In view of the prospect of abundant cheap electric power becoming available, the Department of Industry & Commerce wrote to the GSR regarding adoption of electric traction on sections of their system.[113] J.R. Bazin, the Chief Mechanical

---

110 Paper No.498, I.Loco.E. Journal No.219, Vol. XLIV, 1951
111 D&SER Board Minute No.25430, 18th July 1922
112 The house of representatives of the Irish Parliament
113 Letter from Dept. of Industry & Commerce to G.S.R., 26th March 1925

Engineer, was requested to prepare a memorandum on the matter for submission to the Government. Bazin enlisted the help of his chief assistant, W.H. Morton, and their report confirmed the view that neither the Irish railway network in its entirety, nor any of the larger railway companies, had the traffic volumes to warrant electrification. However, they were of the opinion that in regard to density of traffic, the 14½-mile line between Amiens Street and Bray was most likely to satisfy the economic limits.

With all of this activity concerning availability of electricity, it is no surprise to find that H.E. O'Brien was corresponding with his old friend W.H. Morton on the subject. On 18th May 1927 he wrote to Morton suggesting that Merz & McLellan might be engaged as consultants for a study of possibilities for electrification of GSR suburban lines. The previous evening he had taken part in a discussion on *Rural Distribution of Electricity in the Irish Free State* [114] at a meeting of the Irish Centre of the IEE held in the Physics laboratory at Trinity College Dublin (TCD). In the meantime, James J. Drumm, a research scientist at University College Dublin, had discovered an electrochemical process that enabled him to invent a new kind of zinc-alkaline cell. This cell was capable of being charged and discharged at a rapid rate on account of its very low internal resistance, while its energy capacity per unit of weight was high. In early 1928, the Executive Council of the Irish Free State Government decided to support the development of the invention, granting Patrick McGilligan, the Minister for Industry & Commerce, authority to use up to £10,500; a company named Celia Ltd. being formed in June 1928 for the purpose. Secrecy surrounded the project, and the first inkling that the public got about the invention was when the Minister for Industry & Commerce made an announcement in the Dáil in late February 1929.

For the next stage in the development of the battery McGilligan required a proper experiment to be carried out under actual working conditions. He invited Sir Walter Nugent, Chairman, and W.H. Morton, Chief Mechanical Engineer, of the GSR to attend at his offices on 8th June 1929 and meet with himself and James M. Fay, joint managing director of Celia Ltd. At that meeting it was agreed that a representative of the GSR, together with an expert on railway electrification, should be given the opportunity of testing the charge and discharge characteristics of the battery under laboratory conditions. Subject to a satisfactory report by their representatives, the GSR undertook to provide a suitable motor coach and charging apparatus so as to enable the battery to be tested under actual railway operating conditions. Laboratory testing of the Drumm battery took place towards the end of August and was attended by W.H. Morton and his Electrical Assistant, Warren Storey, together with the 'expert on railway electrification' — H.E. O'Brien. The

---

114 Paper presented to the Irish Centre of the IEE on 28th April 1927 by Sean McEntee, Minister for Finance, who was trained as an Electrical Engineer.

battery that was the subject of the tests consisted of two electric cells only and the testing was carried out using the usual electrical measuring instruments. The tests demonstrated the capacity of the battery to receive a charge of electricity at a high rate in a short time and to discharge the same efficiently in a prescribed period.

On the basis of this outcome it was decided to convert one of the two petrol-engined Drewry railcars that had been purchased by the GSR in 1928, which was considered suitable for the purpose of the first railway test of the Drumm battery. No.386 was available and out of traffic use at the time, and the necessary arrangements were made to equip it with electric motors and control equipment. By mid-June McGilligan was anxiously awaiting the result of trials in order to make a pronouncement in the Dáil. Trial runs actually commenced on Monday 28th July, but the first 'out road' trial to be reported in the press was that which took place on 30th July 1930. On the very next day Eoghan O'Brien was one of nine people invited to board the prototype Drumm battery railcar with W.H. Morton, Warren Storey; and Chief Draughtsman, W.J. Ash for a trial run from Inchicore to Hazlehatch and back.

Prototype Drumm battery railcar converted from GSR Drewry petrol railcar No.386 in 1930 - [GSR © IRRS collection]

So much interest had been aroused by press reports of tests with the prototype Drumm battery railcar that H.E. O'Brien wrote to the *Irish Times* on 15th November 1930 as he felt it desirable to give the general public some idea of the problems involved in its development. Writing under the nom-de-plume "Engineer", he first pointed out three important factors that had to be considered in comparing the Drumm battery with the best existing types, namely cost, weight and durability. Stating that lead-acid batteries had a durable life of from five to ten years, he pointed out that some years must elapse after the first really large Drumm batteries were put into service before any definite opinion could be formed as to their actual durability. He then went on to make a very important point relative to the capacity of the Drumm battery to rapidly take a large charge noting that, unless a very large number of these batteries were in use and being charged at different times of the day, it would throw random excessive loads on the generating station supplying the charging current and on the plant used for converting the alternating current supply into the direct current required for charging the battery. He concluded by suggesting that if trains equipped with Drumm batteries were to be built for railway service in the near future at least one train should be equipped with lead-acid batteries as a basis for an exact comparison, and he stressed that unless this was done the experiment would be of little value.

The preliminary trials having proved satisfactory, it was recommended that two trains, each consisting of two bogie carriages about 60 feet long, should be constructed as a further step in the development of railway traction based upon the Drumm battery system. These were to be suitable for service either separately or coupled together on the Amiens Street – Bray and Harcourt Street – Greystones lines of the DSE section of the GSR. On the 26th November 1930 McGilligan secured a vote in the Dáil for an additional expenditure of £25,000 for the two double coaches, each capable of seating about 100 passengers and weighing about 75 tons, including the battery. The design was to be such that each train would achieve an acceleration rate of about 1 mile/hr/s and a maximum speed of between 50 and 60 mile/hr. It was felt that the battery trains would give all the advantages of third rail electrification, but at a much-reduced cost.

Two-coach articulated Drumm battery train 'A' passing Killiney on trial in late 1931 with Mount Eagle in the background at the right

Construction of the two articulated units was put in hand at Inchicore, but it was to take until the end of November 1931 before the first set, Unit 'A', was ready for trial runs. The official trial went ahead as scheduled on 2nd December, the train departing from Westland Row for Bray at 11.0 a.m. with W.T. Cosgrave, President of the Irish Free State Executive, on board. Many political, diplomatic and railway dignitaries were on the train, and Eoghan was amongst the invited guests. Unit 'A' went into service on 13th February 1932, but the second set, Unit 'B', was not completed until August of that year. The batteries from these two units were rebuilt to an improved design in 1938 and the trains continued in service with the GSR and CIE until 1948.

During the early 1930s Eoghan also worked for an American millionaire (possibly the explorer Lincoln Ellsworth), who wanted to establish a daily steamer service between Montauk Point, New York, and Le Havre in France, doing the trip in four days and flying aeroplanes on and off the ships. Eoghan also tried to purchase the R100 airship for him. In 1924, the British Government undertook to establish communication links by air with the far corners of the Empire. It was decided to construct two entirely new airships to serve routes to Montreal, and Karachi with a view to reaching Australia. The Airship Guarantee Company, a subsidiary of Vickers, won the contract to design and build one ship, the R100, which was assembled at Howden, Yorkshire; and the R101 was built at Cardington by a Government-sponsored team. For R100 Barnes Wallis, assisted by Neville Shute-Norway as his chief calculator, used new design techniques to produce a unique and efficient craft. R100 departed for Canada on 29th July 1930, covering 3,364 miles in 78 hours 49 minutes. During its twelve-day stop in Montreal it also made a 24-hour passenger trip via Ottawa, Toronto, and Niagara Falls. R100 left Canada on its return flight on 13th August and, making use of thermal effects from the Gulf Stream, made the 2,995 mile journey in 57 hours 56 minutes, arriving on 16th August 1930. The next day it was put back in the hanger and the crew switched their attention to the first long-haul flight of R101.

On the night of 4th October R101 left on its maiden flight for the Imperial Conference in India with the Secretary of State for Air, Lord Thompson of Cardington, and most of the design team on board. Not long into its flight, at 2.08 a.m. on 5th October 1930, R101 struck the ground near Beauvais, France, and was destroyed by fire, killing 48 people including Lord Thompson. The loss of the R101 bought to an end the Imperial Airship project and, following an official inquiry into the disaster, the decision was taken to abandon all future flights. R100 was seen as very advanced for its time, so much so that the Americans offered cheap or even free helium to inflate it in return for technical information. But all of this came to naught, and even Eoghan O'Brien's attempt to conclude a deal for its purchase failed, and so R100 was scrapped in 1931.

It was as far back as 1751-52 that the first proposals for a Channel Tunnel were put forward, but it was not until 1833 that the first systematic survey of the Channel was undertaken. In 1866-67 various schemes for tunnels were proposed and by

1868 a preferred scheme was put before the British Channel Tunnel Committee, which gave authority for two pilot tunnels. Work on this scheme started in 1881 and a 2,020-yard tunnel was bored from Shakespeare Cliff under the sea towards Dover harbour. Military opposition to the tunnel in England resulted in construction being halted in 1883. In 1906 the Channel Tunnel Company and l'Association du Chemin de Fer sous-marin entre France et l'Angleterre proposed a new scheme comprising two 20ft diameter tunnels suitable for electric railway operation. A new report on the geology was published in 1919 and a 12ft. diameter trial tunnel 419ft. long was bored in 1923.

Soon after winning the election in 1924, Britain's first labour party Prime Minister, Ramsay MacDonald, took an interest in the project. Sir William Bull[115], Conservative MP for Hammersmith and Chairman of the Channel Tunnel Committee, secured a commitment from him to review the question of the tunnel without delay and report on the same to Parliament. MacDonald consulted the Committee of Imperial Defence, which was strongly opposed to the idea of a channel tunnel, and on 7th July 1924 he announced that the Government would not support the scheme.

By early 1929 the conservatives were in government again and on 1st March Richard Gardiner Casey, the Australian Political Liaison Officer in London, who had studied engineering at the Universities of Melbourne and Cambridge, sent a note to the Prime Minister, Stanley Baldwin. In it he noted that Baron d'Erlanger and Sir William Bull were the moving spirits in the Channel Tunnel Company, and he suggested that that they were not the sort of people who would take any standpoint other than that of their own pockets. Also in March 1929 Sir Robert Thomas, Conservative MP for Anglesey, asked the Prime Minister if he would cause the objections of the Committee for Imperial Defence to the construction of a Channel Tunnel to be set forth in detail, to which Baldwin replied in the negative.

On 26th March, in reply to Sir William Bull, Baldwin stated, "the best plan for starting an investigation will be to set up, under the auspices of the Committee of Civil Research, an impartial inquiry into the economic aspects of the proposed tunnel and other projects for cross-channel communications". Consideration of the political and military aspects would be postponed until the report had been received. By late October the Committee, under the chairmanship of Sir Edward Peacock, produced a majority report stating that the tunnel would be "of economic advantage to this country", and recommending that it should be constructed and maintained by private enterprise. By that time, MacDonald was back in power and once again he invoked the assistance of the Committee of Imperial Defence with predictable results.

Meanwhile, in late 1928 a proposal was published for a broad-gauge railway

---

115 Sir William Bull died in 1931. His son, Anthony, was Vice-Chairman of London Transport 1965-69.

CHAPTER THIRTEEN **Retirement at Killiney**

between the British and French capitals. In the words of C.F. Dendy Marshall, "the scheme was worthy of Brunel at his best". Brunel's 7ft 0¼in. gauge was chosen and the proposal was to run between London and Paris in 2h 45min at an average speed of 92 mph. Speed in the 44-mile tunnel under the Channel was to be kept down to 60 mph. At that stage the design of the locomotives for the London & Paris Railway (L&PR) had not been settled, but drawings had been produced for the passenger coaches. The scheme envisaged maximum axle loads of 20 tons, and the bogies for the passenger cars were to be centred by inclined planes and not by swing links. Dendy Marshall pointed out that it was difficult to choose the best angle for inclined planes, noting that they were impossible to adjust and that their operation would be affected by the state of lubrication. He suggested that spring control would be a better solution.[116]

For an estimate of the detailed working expenses of the L&PR, William Collard, the representative of London & Paris Railway Promotions Ltd. of Saville Row, London, was put in touch with Col. O'Brien in regard to electric traction aspects. Collard made a visit to Dublin in October 1932, following which Eoghan wrote to him suggesting the use of data from Francis Lydall's then recent paper read to the British Association meeting in York.[117] In January 1933, Collard was able to send a copy of the revised estimates of working expenses to O'Brien for his consideration and endorsement.

In regard to freight traffic reference was made to the coal trains operated by the LMS between Toton and London. On 5th December 1932 Collard had written to Sir Henry Fowler seeking information and was informed that the running of coal trains of 1,300 tons from Toton to London had recently been accelerated and that the trains were hauled by a single 2-6-0+0-6-2 Garratt type locomotive having a tractive effort of 45,620 lb @ 85% boiler pressure. Collard requested H.E. O'Brien to convert this tractive effort figure into horsepower for inclusion in the estimates. His calculations survive, handwritten on the reverse of Collard's letter, and they show that the power required at various running speeds would be as follows: 10mph = 1,100hp; 20 mph = 1,333hp; 30mph = 1,440hp; and 40mph = 1,560hp.

In January 1933 Collard wrote to Baron Emile d'Erlanger, Chairman of the Channel Tunnel Company, and summarised the problems for the tunnel project as follows:

1. The announcement by the Southern Railway to apply to Parliament to construct and operate a train ferry between England and the Continent was seen as opposition and, if in existence, the prospects of a Channel tunnel would receive an indefinite set back.

2. The Southern Railway's objection to the Channel Tunnel, as demonstrated by late Chairman's speech at the House of Commons meeting.

---

116 *The Proposed London to Paris Railway*, Railway Engineer, March 1929
117 Lydall, F.: *Electric Traction*, paper read at British Association Meeting, York, 5th September 1932

3. A strong Committee appointed by Stanley Baldwin had reported that the Channel Tunnel would be of economic advantage to the United Kingdom. However, his successor, Ramsay MacDonald was opposed to the tunnel.

4. MacDonald's personal and active intervention in the debate, which had resulted in Ernest Thurtle's motion in favour of the Channel Tunnel being lost by a mere 7 votes.[118]

5. The Channel Tunnel scheme needed to be reinforced by amalgamation with the L&PR scheme as a link in the direct railway in order to capture the imagination of men sufficiently to over-come the opposition of the Prime Minister.

The L&PR proposal for a new direct railway between the two capitals was reflected by the Channel Tunnel Committee in the following terms "this scheme seems to us prohibitive in cost and impracticable in the present state of engineering knowledge and experience." However, Metropolitan-Vickers, who had successfully designed the locomotives for the electrified service on the Bombay-Poona section of the Great Indian Peninsula Railway[119], refuted the charge that the scheme was impracticable. Here, Collard pointed out that, working to the specifications of their consulting engineer for electric traction (H.E. O'Brien), Metropolitan-Vickers had prepared a complete set of drawings for the L&PR express locomotives and had stated in their report that "The journey between London and Paris can be accomplished in the time required (2hr 45m) and we feel certain that the locomotives have ample capacity."

Interestingly, in relation to air resistance, a 'parabolic wedge' design was proposed in 1929 for the front of the locomotives for the L&PR, foreshadowing by some six-years the emergence of Gresley's 'A4' design for the L&NER. It was estimated that 3,712hp was required for 120mph in calm conditions but that this would rise under head-wind conditions. Whilst on the subject of Gresley's streamlined pacific locomotives, it must be mentioned that Eoghan's first cousin, Sir Murrough Wilson, had the honour on 8th April 1939 of having one of Gresley's 'A4' class locomotives, No.4499 (a locomotive of the same class as the famous No.4468 *Mallard*) named after him in recognition of his services to the company.

Sir Murrough Wilson, Deputy Chairman, L&NER - [Reproduced from Railway Gazette May 10, 1946, by permission of the Editor]

Murrough John Wilson had been appointed to the Board of the NER on 19th April 1912 and had served with the Yorkshire Regiment during World War 1. In January 1923 he was selected as one of the NER representatives

---
118 Pro-tunnel motion put before the House of Commons on 30th June 1930.

119 Lydall, F.: *The Electrification of the Suburban and certain Main-line sections of the Great Indian Peninsular Railway*, I.E.E. Journal, Vol.71, December 1932

CHAPTER THIRTEEN **Retirement at Killiney**

to serve on the L&NER Board and he was created a KBE in the King's birthday honours of June 1927.[120] His experience with the Navy, Army & Air Force Institutes suited him well when he was appointed to follow Col. W.J. Galloway as chairman of the Hotels Committee. His most useful period in office came when, in September 1933, the L&NER appointed him to succeed Lord Grey of Fallodon as chairman of the Traffic Committee and chairman of the North Eastern Area Board. He was thoroughly happy in dealing with proposals for improving railway working and, when on officers' tours of inspection, he delighted in discussing operating questions with the staff that were handling traffic. He succeeded the late Lord Farringdon as deputy chairman of the L&NER on 27th April 1934; a position that he was to hold until his death on 30th April 1946. At the time of his death he represented the L&NER on the Cheshire Lines Committee and on the Board of the Forth Bridge Railway Company. He was also a Director of the London Electricity Supply Corporation Ltd. and other companies.

*Official photograph of A4 Class locomotive 4499 **Sir Murrough Wilson**, with new nameplates affixed after its re-naming from Pochard - [Doncaster 39/35 © NRM, York]*

During R.E.L. Maunsell's presidency of the I.Loco.E. in 1928-29, H.E. O'Brien was elevated to an Honorary Member of that professional body. This notable event was reported at the 18th AGM held at Denison House, Vauxhall Bridge Road, London, on Thursday 25th April 1929. Following his return to Killiney Eoghan was also an active participant in the affairs of the Irish Centre of the I.E.E., serving as a member of the local Committee. At the Committee meeting held on 24th April 1930 he was elected vice-chairman of the Irish Centre for the 1930-31 session commencing on 1st October. Discussions on papers to be presented during the session took place at the Committee meeting held on 30th October, and H.E. O'Brien agreed to give a paper on 15th January 1931 if no alternative was available. The latter was indeed the case, and the text of his unpublished paper *Electric Traction*, which has survived in the R.N. Clements collection, now appears in print for the first time in Appendix Eight by kind permission of the Irish Railway Record Society.

120 *London Gazette*, 3rd June 1927

*L&NER A4 class locomotive No.4499 **Sir Murrough Wilson** approaching York from the north with an Up train in the summer of 1939.- [E.R.Morten]*

In the absence of F.H. Whysall, H.E. O'Brien took the chair for a paper on *Heavy Duty Rectifiers and their application to Traction Sub-stations* read by J.W. Rissik at a meeting of the Irish Centre of the I.E.E. held in the Physics laboratory at TCD on 20th April 1931, an opportunity that was to occur for him again on 4th May 1931. In the meantime, at the Committee Meeting of the Irish Centre, held on 20th April 1931, he was elected chairman for the 1931-32 session. In his chairman's address, delivered to the Irish Centre of the I.E.E, in Dublin on 29th October 1931, he directed his remarks principally toward the younger members. Having opened with a brief review of the progress that had been made in electrical engineering he then turned to his main theme — the importance of vision and sound education in equipping young engineers for their profession. Placing emphasis on the constant endeavour made by American business schools to introduce their students to the practical realities of business in close co-operation with the surrounding business community, he stated that: "No one can visit these large American schools of business without feeling some misgivings on the industrial future of any country which lacks them." In his address Eoghan also pointed out that while the young engineer must depend on his own intelligence and strength, he must also look for a third qualification: experience. This, he said, could be acquired in two directions only; future time and from elders in the profession.

As Chairman of the Irish Centre he attended the 4th Annual Dinner of the association of University Electrical Engineers held at the Shelbourne Hotel, Dublin, on 6th February 1932. The guest speaker was Patrick McGilligan, Minister for Industry & Commerce, who talked on the Shannon Scheme, stating that it was the greatest gamble the country had ever undertaken. During his speech, Eoghan O'Brien expressed his strongly held opinion that all engineers should spend some time in the workshops as it was there that one got to know his fellow men. He also advised young engineers to leave their country for a short while at some stage during their careers as this would enable them to gain a perspective which they would not get at home. Another important event took place on 12th May 1932, when Col. & Mrs O'Brien, together with the Committee of the Irish Centre of the I.E.E., held a reception followed by a dance at the Royal Hibernian Hotel, Dawson Street, for Mr J. Donaldson, President of the I.E.E., on the occasion of his visit to Ireland.

In May 1933 H.E. O'Brien read his paper entitled *The growth of Electricity Supply and its relationship to civilisation*, before the Statistical & Social Inquiry Society of Ireland. In his home country recognition of his professional status as an engineer came about on 8th January 1934 when he was elected a Member of the Institution of Civil Engineers of Ireland. His chairman's address to the Irish Centre of the I.E.E. had shown that he had not lost the 'Horwich' approach to the all-round development of young engineers, and circumstances were soon to place him in a position to share his wealth of knowledge and experience with students of the profession. In August 1935 George Marshall Harriss retired from the position of General Manager of the DUTC. He had not been enjoying the best of health and, having reached the age of 70, he also decided to give notice to TCD of his intention to resign the post of Lecturer in Electric Traction at the end of the year.[121] The vacancy created by his resignation was filled by the Board of TCD with the appointment of Eoghan O'Brien to the position of Lecturer in Electric Traction on 11th January 1936.[122]

In the summer of 1936 during his annual visit to Switzerland, and when on a solitary walk of many hours in the Alps, it occurred to Eoghan that there would never be a more suitable opportunity to carry out the electrification of the Dublin – Bray line. This thought was prompted by correspondence he was having with William Arter regarding lightweight high-speed trains. Arter had been his former assistant on the Liverpool – Southport line and was, by that time, head of the railway department of the Allis Chalmers Company. On his return home, Eoghan put pen to paper and sent details of his scheme to his old friend W.H. Morton in October 1936. In his letter he suggested that if he could induce C.H. Merz to interest himself in the scheme they would form an Irish firm of consulting engineers, O'Brien, Merz & Partners. He felt that the name Eoghan O'Brien would strike the right patriotic note, particularly if the relationship as a grand nephew of William Smith O'Brien was emphasised and the 'Colonel' was discreetly dropped. Furthermore, he was of the opinion that the Merz name was sufficiently German sounding to satisfy any anti-British bias and that his inclusion would give the benefit of the worldwide experience of Merz in railway electrification. He proposed that some other Irish engineering entrepreneurs could be associated with the firm in order to make it 80% Irish-owned.

Eoghan went on to suggest that the GSR might employ such a firm to strengthen

---

121 The post of Lectureship in Electric Traction at Trinity College was founded in 1903; the title of the post being changed in 1937 to that of Lecturer of Railway Traction.
122 TCD MUN/V/5/25 — Register of the meetings of the Board of Trinity College Dublin, 1934-42.

their position in regard to any proposals they would wish to put forward. In regard to technical features he proposed electrification at 3,000-V DC, with a single sub-station at Dun Laoghaire. The overhead structure would be constructed from materials that would not corrode, such as stainless steel and phosphor bronze, thus considerably reducing maintenance in spite of the salty atmosphere to which this coastal line is exposed. He recommended the use of very light trains of the type developed by the Budd Company in America, which he considered should result in 40% less energy consumption and 30% less in repairs than on the average electrified railway. Fully automatic signalling was envisaged, except for the station layouts at Dun Laoghaire and Bray, and new halts were proposed at Vico Road and Shankill.

In conclusion, he noted that the correct procedure in contracting for the electrification would be to let every nationality tender and to accept the lowest tender that met the specification. He felt that some slight bias might be permitted towards Lancashire and America on account of the large population of Irish birth and descent in both places! He admitted to Morton that he had always longed to have a finger in the electrification of what he termed his 'home railway', and confessed that he would not be averse to earning a little money, but he believed that the work itself would be the major interest. He asked Morton to give the proposition some thought and he invited him to come out to *Mount Eagle* and talk it over, if it was of interest.

In his reply, W.H. Morton stressed that some of his Directors had left him in no doubt as to their strong opposition to any scheme of electrification involving heavy capital expenditure. Despite their view, Morton believed that electrification of the Dublin-Bray line must be undertaken eventually. In conclusion he pointed out that the Free State Government had employed Merz & McLellan in connection with the Drumm battery experiments, and he confided that he had always been puzzled at the nature of their Report considering their worldwide experience in the installation of continuous current contact systems of railway electrification.

With the upsurge of interest in alternatives to the reciprocating steam engine, which came about following the successful introduction of W.A. Stanier's *Turbomotive* by the LMS in June 1935, correspondence abounded in the technical press for many months. In a letter to *Modern Transport*, published in the edition of 27th March 1937 under the *nom de plume* "A Former Locomotive Engineer", H.E. O'Brien pointed out that it hardly seemed possible that the prime cost of a turbine, turbo-electric or diesel-electric locomotive could compete with that of the modern three-cylinder locomotive. He felt that in order to make a true comparison between the different types it was necessary to draw up complete statements of the total costs of operation, including interest on capital cost, depreciation and obsolescence.

He went on to state that it seemed probable that a new design could cost anything from 25% to 100% more than a conventional design, and that fuel economy alone was not the dominant factor unless the new design delivered other advantages in reduced operating wages and repair costs. He felt that the bulk of locomotive stock, which had to be handled by all sorts of drivers and shed staff, should be of the simplest possible construction and highly standardised. The determination of the economic case for a new type of machine was, he considered, a very complicated matter that could be obtained only by the closest co-operation between the engineer and the accountant.

The LMS Turbomotive No.6202 upon which Col. O'Brien commented in March 1937 - [LMS Official]

Having pointed out the factors that must be considered in opting for a new design, O'Brien concluded by stressing the need for constant and persistent experimental work and cited the achievements of Sir John Aspinall in introducing modern methods of standardisation, operating and accounting in the railway business. In expressing his view that there should be no grudging of money spent on promising experiments he recognised that it was often difficult for the engineer and accountant to agree on the scale of experimentation. He added that his own experience was that much money was wasted by small-scale experimentation that lacked a definite aim, and which was insufficiently supervised and inadequately recorded.

Among the fashionable events in the Dublin social calendar during the 1930s were the garden parties that Frances O'Brien organised each year at *Mount Eagle* in order to raise funds for charity. These occasions were always well attended, up to as many as 200 guests being known to visit *Mount Eagle* in an afternoon to view her beautifully tended garden with its great variety of flowers and to gaze upon the vista of Killiney Bay; tea being served in the house. On the personal front, a happy family event occurred in 1935 when it was announced that Lt. Brian Eoghan O'Brien RN and Elizabeth, only daughter of Dr & Mrs S.S. Strahan of Hong Kong, had become engaged; they were married in Hong Kong on 10th February 1936. The happy couple managed to get to Ireland in April in time for the 1936 garden party at *Mount Eagle*, at which event they were presented to friends and relatives. There was more joy for the family with the arrival of a granddaughter, Olivia Fiona, on 22nd November 1938. Although the clouds of war were once more gathering over Europe, the enjoyment of the last few years of peace were capped for H.E. O'Brien on 28th June 1939 when an M.A. degree was conferred on him by TCD in recognition of his professional status and position on the staff of the college.

The GSR put two more two-coach articulated Drumm trains, Units 'C' & 'D', into service in September and October 1939, and in a personal letter to W.H. Morton, written on 11th November, Eoghan commented on the fact that all four Drumm trains were working exclusively on the Harcourt Street line. He pointed out that the transfer of the old units away from the Amiens Street – Bray route, together with the increased frequency on that route, was resulting in the houses and gardens along that line being exposed to a greater degree of dirt from locomotive smoke. As the coastal route was far more residential than the Harcourt Street line, he suggested that steam locomotives could be maintained on the latter and that the Drumm trains might be put back on the Amiens Street – Bray line. In his opinion neither line was suitable for battery operated trains, particularly of the Drumm type, and he remarked that it would be very interesting to see how Warren Storey produced calculations that could satisfy anybody! The only section that he considered anyway suitable for the heavy Drumm battery trains was the six-mile level stretch from Westland Row to Dun Laoghaire.

Once again, H.E. O'Brien expressed his view that electrification of the Amiens Street – Bray line was the proper solution, particularly when the cost of coal and wages and the very low cost of electricity in Ireland were taken into account. He felt that, because the heaviest traffic on the line occurred during the summer months, a particularly advantageous arrangement might be made for supply of electrical power. Returning to the question of the dirt caused by the steam locomotives in a postscript to his letter, he pointed out that his wife's hands got so dirty whenever she picked flowers, whereas they used to be clean. He added, however, that he had no complaint to make in regard to bad firing of the steam locomotives!

Replying on the 7th December, W.H. Morton regretted that Mrs O'Brien had to complain about 'deposited carbon', adding that he knew from sad experience how distasteful it was and he apologised for not preserving her flowers in their "pristine beauty and cleanliness". In addressing the technical points raised by H.E. O'Brien, Morton stated that the electric battery trains had been "tried and found wanting" on the Amiens Street – Bray section. Although the line is level between Westland Row and Dun Laoghaire, they could not stand up to the additional energy required to haul extra coaches at peak hours, which occurred twice a day and always at holiday times and at weekends during the summer months. This was the reason that the GSR had finally and definitely decided to transfer them all onto the Harcourt Street line.

CHAPTER THIRTEEN Retirement at Killiney

In his letter, W.H. Morton noted that the various factors and economics of suburban railway electrification had been under consideration by the GSR periodically since 1925. He added that he himself had advocated the electrification of the Amiens Street – Bray section on three separate occasions, in 1925, 1933 and 1936, but that lack of capital resources had prevented the schemes from going ahead. He concluded by providing Col. O'Brien with the various estimates that had been produced, as follows:

1925	1,200 Volt D.C. Overhead System	£285,000
1933	1,500 Volt D.C. Overhead System	£363,700
1936	650 Volt D.C. Third Rail System	£322,000
1936	1,500 Volt D.C. Overhead System	£310,000

Tragedy was to strike both the O'Brien and Merz families during the Second World War. On 10th January 1939 Lt-Cdr. (Engineer) Brian Eoghan O'Brien had joined the crew of H.M. Submarine *Thames*, under the command of Lt-Cdr. W.D. Dunkerley. She sailed from Dundee on 22nd July 1940, but was reported overdue on 3rd August 1940. The officially accepted theory is that she hit a mine in a newly laid German minefield at approximately 57°N, 3°E. Lt-Cdr. W.D. Dunkerley, Lt-Cdr.(E) Brian Eoghan O'Brien and three other officers (Lt. F.R.C. Talbot, Lt. F.H. Morris and Lt.(RNR) D.E.T. Newell) together with 56 ratings were lost with the *Thames*.

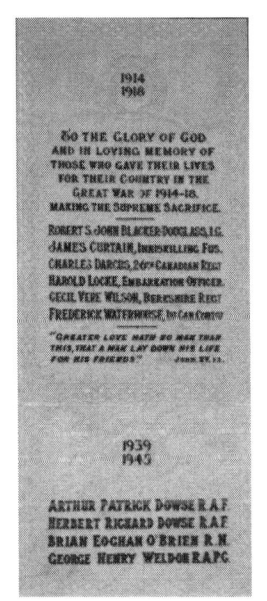

War Memorial in Holy Trinity Church, Killiney, Co. Dublin. - [Author]

Charles Hesterman Merz - [Reproduced from Railway Gazette October 25, 1940, by permission of the Editor]

203

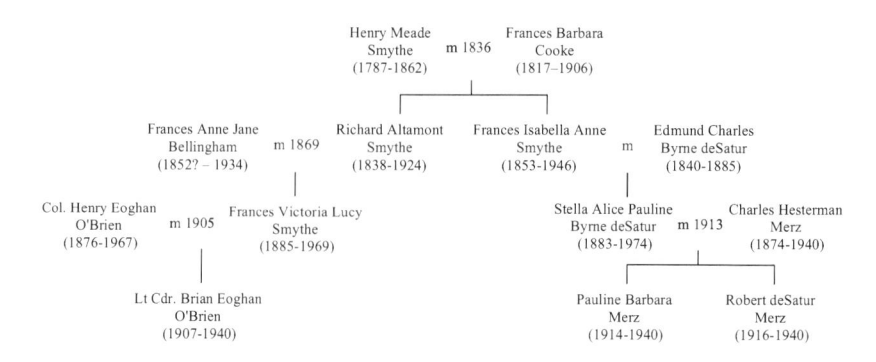

```
                    Henry Meade            Frances Barbara
                    Smythe       m 1836    Cooke
                    (1787-1862)            (1817-1906)
                              |
   ┌──────────────────────────┼──────────────────────────┐
Frances Anne Jane        Richard Altamont    Frances Isabella Anne   Edmund Charles
Bellingham    m 1869     Smythe              Smythe          m       Byrne deSatur
(1852? – 1934)           (1838-1924)         (1853-1946)             (1840-1885)
         |                                            |
Col. Henry Eoghan    Frances Victoria Lucy       Stella Alice Pauline    Charles Hesterman
O'Brien      m 1905  Smythe                      Byrne deSatur  m 1913   Merz
(1876-1967)          (1885-1969)                 (1883-1974)             (1874-1940)
         |                                              |
   Lt Cdr. Brian Eoghan                     ┌───────────┴───────────┐
   O'Brien                              Pauline Barbara        Robert deSatur
   (1907-1940)                          Merz                   Merz
                                        (1914-1940)            (1916-1940)
```

Family tree showing the relationship between H.E. O'Brien and C.H. Merz

A little over two months after the loss of the *Thames*, Charles Hesterman Merz and his children, Pauline Barbara and Robert deSatur, were killed when a German bomb hit the Merz home at 14 Melbury Road, Kensington, during an air raid on the night of 15th October 1940. The deaths of such a close friend and family members, coming so soon after the loss of their own son, were a further sad blow for Eoghan and Frances. Not long after these events, Eoghan wrote to President Douglas Hyde in December 1940 expressing concern that, although the Hospital Sweepstakes Trustees held some £8½ million, no additional hospital beds had been provided in Dublin in the ten years since the Sweeps commenced.[123]

The official reaction was, to say the least, dramatic; Frank Aiken, the Minister for the Co-ordination of Defensive Measures, directed Thomas J Coyne, Assistant Controller of Censorship, to summon Col. O'Brien for an interview in connection with certain letters to persons outside the country which had come under observation in the Postal Censorship. When he met with Col. O'Brien on 9th January 1941, Coyne indicated that the letters could be distinguished from legitimate correspondence on personal affairs by the fact that each of them contained matter that was calculated, in a greater or lesser degree, to endanger the security of the State. He stressed that he wanted to make it plain that the State was in no way concerned with Col. O'Brien's private opinions on questions of foreign or domestic politics. However, Coyne stated that it was his duty to warn Col. O'Brien that the authorities took a grave view of the letters which contained matter that, in the opinion of the law officers of the State, was nothing short of sedition. He indicated that, if that sort of correspondence continued, the authorities would be compelled to take action, and specifically mentioned deportation, or internment, or a prosecution before the Special Criminal Court!

---

[123] In January 1973, following months of investigation, the *Sunday Independent* exposed the Irish Hospital's Sweepstakes as a complete scam. It was discovered that Irish hospitals had received less than 10% of the value of the lottery tickets marketed in their name throughout the world for fifty years. The scandal was also described in *The Greatest Bleeding Heart Racket in the World* (Damien Corless, Gill &Macmillan, 2010)

Col. O'Brien replied by protesting that he was merely expressing his private views which he thought he was fully entitled to express in his private correspondence. He denied that the letters to which T.J. Coyne had referred to formed part of a connected series written with the deliberate object of endangering the country. On the contrary, he said that the letters were written sporadically in response to a particular mood which was liable to change with the course of the war, and all of them were written under the impulse of a strong feeling accentuated by the recent death of his son. He said they were quite wrong in thinking that he had any desire to see the British back in the State; on the contrary it was his view that there could be no greater calamity than the return of the British. He added that he was quite willing to give an assurance that he would discontinue that line of correspondence and would go even further and undertake not to mention war at all, or, at any rate, Ireland's position and problems in relation to the war.

There the matter rested until Saturday 12th July 1941 when Col. O'Brien and his wife were among a party of guests for tea at *Áras an Uachtaráin*, at which Sir John Maffey, British Representative to Ireland, and Lady Maffey were also present.[124] The Secretary to the President, Michael McDunphy, was in attendance, and two days later he wrote to T.J. Coyne, noting that he had spoken with the chief Press Censor, and indicating that Col. O'Brien should not be invited again to the Áras without further consultation with him. In his reply Coyne noted:

> "O'Brien is an engineer who served in the British Army during the last war with some distinction. He is an authority on electric traction and was employed at one time on one of the railways in England. He is the son of a former Irish Land Commissioner, an old Etonian, and one of the claimants to a lineal descendant of Brian Boru. He bears the reputation, in the circles which he frequents, of being a decent fellow, and I have been told that he is a man of considerable ability in his own line."

He added that in correspondence

> "The Colonel did not confine himself to providing material for parliamentary questions, but urged his correspondents, in both England and America, to use whatever influence they possessed to awaken the British to their supposed danger and, if possible, to induce the Americans to intervene."

In a file note, dated 11th August 1941, Michael McDunphy recorded the following;

> "in his letters Col. O'Brien referred to what he described as 'the shameful neutrality' of Ireland... ... made slighting references to the character of the Irish people and, showed that his actively hostile attitude towards his own country had not changed since his interview with Mr Coyne last January."

> "This new development confirms my view that Col. O'Brien is, from the national point of view, highly objectionable and unfit to be a guest in the President's house."

---

124 *Áras an Uachtaráin*, the official residence of the President of Ireland situated in the Phoenix Park, was formerly the Viceregal Lodge, was the residence of the Lord Lieutenants who oversaw British rule in Ireland until 1922, and from then until 1936 it was the residence of the Governors General (the Crown representatives in the Irish Free State).

However, T.J. Coyne took no further action.[125] It is to be regretted that, apart from that initial letter to the President in December 1940, we do not have any of Col. O'Brien's correspondence by which to assess the veracity of McDunphy's notes. It is not possible to determine whether McDunphy's opinion of him was based solely on the letter to the President and/or the meeting at *Áras an Uachtaráin*. What is known is that Michael McDunphy has been described as "a controversial, outspoken and temperamental civil servant", and that on the death of Adolf Hitler in April 1945, he called on Dr Hempel, German Representative in Ireland, to express condolences on behalf of the President. This may indicate where McDunphy's loyalties lay — whilst Eoghan O'Brien, although he had been educated at Eton and had spent much of his career in Great Britain was, and always claimed to be, Irish.

H.E. O'Brien continued to hold the post of Lecturer of Railway Traction at TCD until the end of the academic year 1953-54 and he was for a time also the vice-president of the TCD Engineering Society. During his time as a lecturer at TCD he always made a point of ensuring that his students were exposed to the practical realities of the profession. The text of his letter dated 4th January 1945 to Edgar Bredin, the then General Manager of CIE, gives an insight to the approach he adopted. In it he wrote, "Before I start my lectures at Trinity I want to show my students a locomotive and a boiler, also a tramway motor and a tramcar controller", to which Bredin replied on 9th January that it would be necessary to make two separate visits "as the locomotive will be at Inchicore and the tramway material at Ballsbridge". H.E. O'Brien was then able to promptly arrange with Professor Passer the exact days on which he could have the students for two separate visits during the week of 22nd to 27th January.

Bed ridden with arthritis in his last days, H.E. O'Brien still always welcomed visitors to *Mount Eagle*, especially if they came to talk about railways. It seems fitting that one of the 20th century's more important railway engineers, who had been born within earshot of trains running along the coastal line between Dublin and Bray, should pass from this life also in close proximity to that same railway line, but to the sound of diesel, rather than steam trains. Henry Eoghan O'Brien died on 9th September 1967 at *Mount Eagle*, Killiney in his 92nd year. His remains were cremated at Liverpool Crematorium at 12:40 on Thursday 15th September 1967. Frances Victoria Lucy O'Brien survived her husband by a further seventeen months. She died at a Dublin nursing home on 15th February 1969 and she was buried at Deansgrange cemetery on 18th February 1969. Following their deaths *Mount Eagle* was put up for sale and Dun Laoghaire Corporation tried to buy the property in order to extend the public park at the Vico Road, but were outbid at auction by Con Smith, a private buyer.

---

125 On his retirement in February 1961, *The Irish Times* described T.J. Coyne as "always one of the most human of civil servants".

CHAPTER THIRTEEN **Retirement at Killiney**

The introduction of electrified services on the West Coast Main Line from Weaver Junction north to Glasgow in May 1974 'flattened' the climb over Shap Fell as envisioned by Col. H.E. O'Brien in 1924. - [Courtesy of Rail Express]

Electrification of H.E. O'Brien's 'home railway' – the first Dublin Area Rapid Transit service passes Mount Eagle on 23rd July 1984 - [David Carse]

So what can one say of this outstanding engineer of the 'Horwich School' whose own vision of the future for railway traction was, to quote an expression from his address to the Irish Centre of the I.E.E., "before his time"? There is no doubt that if the LMS had adopted his ideas for electrification of the Crewe-Carlisle section it could have been carried out for less than it eventually cost. However, even in old age Eoghan recognised that his system would have become out-of-date and needed renewal. The final thought must rest with the man himself who, during an interview just before he died, while reminiscing that his trials had proved that for a 1,000-ton train with a 3,000 hp electric locomotive "Shap summit could be virtually ironed out", pointed out that the then recently completed Euston to Manchester and Liverpool electrification would not be a 100% pay off until it was continued along the line to Scotland.[126]

It was the success of that electrification, which had cut journey times by a third and more than doubled passenger numbers and revenue in its first four years of operation, which finally pushed the case for an extension of the 25kV electrification to Glasgow. British Rail inaugurated electric services on the 255-mile route from Weaver Junction to Glasgow on 6th May 1974 — just eight weeks after the 50th anniversary of the first presentation of Col. H.E. O'Brien's paper to the I.E.E. proposing electrification along part of the same route between Crewe and Carlisle.

Although Eoghan did not live to see it, his 'home railway' was eventually electrified at 1,500-V DC with overhead line equipment as part of the Dublin Area Rapid Transit (DART) system, on which CIE commenced running passenger services on 23rd July 1984. Another scheme that he had been associated with also came to fruition when, on 6th May 1994, Queen Elizabeth II and President François Mitterrand officially opened the Channel Tunnel. Although high-speed lines were brought into use on the French side when the Tunnel was opened, the rail network on the English side constrained the London - Paris journey time to 2h 55m. Construction of the dedicated high-speed line through southeast England with 25kV AC overhead electrification — the Channel Tunnel Rail Link (CTRL) —akin to Collard's L&PR scheme, addresses this problem.

Rt.Hon. Tony Blair, the British Prime Minister, opened the first section of the CTRL, from Folkestone to Fawkham Junction in Kent, on 16th September 2003, and this reduced the journey time between London and Paris to 2h 35m. During a special trial run on 4th September 2007 the fastest ever rail journey time between the French and British capitals was completed in a record-breaking 2h 3m. The train left Gare du Nord station in Paris at 10.44 CET and arrived at St. Pancras International at 11.47 BST; travelling at speeds of up to 200 mile/h (320 km/h). The official opening of St. Pancras International and the launching of High Speed 1 (as CTRL is now known) was performed on 6th November by Queen Elizabeth II accompanied by the Duke of Edinburgh, and *Eurostar* services were transferred from Waterloo to St. Pancras International on 14th November 2007. We now definitely live in an age where main line electric railway traction is no longer questioned, and it is hard to comprehend how different it must have been nearly 95 years ago for Eoghan O'Brien to argue the case for his vision of the future for railway traction.

---

126 *Horwich and Westhoughton Journal,* Friday 29th September 1967

Appendix One

L&YR Chief Mechanical Engineer's department senior officers

Chief Mechanical Engineer	Assistant CME (Works Manager Horwich)	Assistant, Carriage & Wagon Superintendent	Head of Outdoor Locomotive Department	Assistant Outdoor Loco Department	Resident Electrical Engineer	Works Manager (Newton Heath)	Assistant Works Manager (Newton Heath)	Assistant (Outdoor C&W Department)
J.A.F. Aspinall 01/10/86 to 30/06/99	H.A. Hoy 04/87 to 30/06/99		C.O. Mackay -/78 to 29/05/01	G. Banks 04/87 to 30/06/99				A.D. Jones 01/08/98 To 30/06/99
		G. Hughes 23/10/95 to 30/06/99						G.H. Roberts 01/07/99 to 30/06/00
H.A. Hoy 01/07/99 to 11/03/04	G. Hughes 01/07/99 to 11/03/04	G. Banks (1/07/99 to 30/09/99		A.D. Jones 01/07/99 to 30/06/01		J. Howarth ?? to 28/02/02	O. Winder 16/08/97 to 30/09/99	H.N. Gresley 01/07/00 to 30/06/01
		O. Winder 01/10/99 to 11/03/04		J.P. Crouch 01/07/01 to 17/02/05			J.P. Crouch 01/10/99 to 30/06/01	C.H. Montgomery 01/07/01 to 11/03/04
		H.N. Gresley 12/03/04 to 17/02/05	A.D. Jones 01/07/01 to 31/03/12			H.N. Gresley 01/03/02 to 11/03/04	H.N. Gresley 01/07/01 to 28/02/02	
G. Hughes 12/03/04 to 31/12/21	O. Winder 12/03/04 to 11/06/09	J.P. Crouch 18/02/05 to 13/06/09		F.W. Attock 18/02/05 to 31/03/12	(Formby) H.E. O'Brien 24/06/03 to 13/06/09	C.H. Montgomery 12/03/04 to 13/06/09	F.E. Gobey 01/03/02 to 31/03/03	F.W. Attock 08/04/04 to 17/02/05
							F.C. Hibberd 01/04/03 to 31/12/07	F.S. Barnes 18/02/05 to 31/12/21
	J.P. Crouch 14/06/09 to 28/02/10	H.E. O'Brien 14/06/09 to 28/02/10			C.H. Montgomery 14/06/09 to 30/01/12	F.E. Gobey 14/06/09 to 28/02/10	R.G. McLaughlin 03/01/08 to 28/02/10	
	H.E. O'Brien 01/03/10 to 31/12/21	F.E. Gobey 01/03/10 to 31/12/21	F.W. Attock 01/04/12 to 31/12/21	H. Housley 01/04/12 to 31/12/21	H. Creagh 31/01/12 to 31/12/21	R.G. McLaughlin 01/03/10 to 30/09/20	H. Creagh 01/03/10 to 30/01/12	
					(Clifton Junction) A. Lund 24/07/15 to 31/12/22	E.F. Merrett 01/10/20 to 31/12/21	A. Lund 31/01/12 to30/09/20	
							E.F. Merrett 01/01/14 to 30/09/20	

Appendix Two

L&YR Electric Stock

Order	Qty	Type	DIA	Length	Width	Seats 1st	Seats 3rd	Builder	Year	Fleet Nos.
A20	28	Third Class Motor Cars	56	60' 4¾"	10' 0"	-	(60) 80	L&YR	1904	3000 - 3027
C20	26	First Class Trailer Cars	57	60' 4¾"	10' 0"	66	-	L&YR	1904	400 - 425
F20	1	Motor Baggage Car	-	54' 0"	8' 9"	-	-	Dick Kerr	1903	3028
L20	1	Motor Baggage Car	59	54' 0"	9' 0"	-	-	L&YR	1904	3029
N21	2	First Class Trailer Cars	57	60' 4¾"	10' 0"	66	-	L&YR	1905	426 & 427
D22	2	Third Class Motor Cars	56	60' 4¾"	10' 0"	-	(60) 80	L&YR	1905	3030 & 3031
M21	2	Third Class Trailer Cars	68	60' 4¾"	10' 0"	-	(80) 98	L&YR	1905	3100 & 3101
R21	10	Third Class Trailer Cars	69	60' 4¾"	10' 0"	-	(80) 92	L&YR	1905	3102 - 3111
K22	2	Third Class Motor Cars	73	60' 4¾"	10' 0"	-	(66) 78	L&YR	1905	3032 & 3033
L22	1	Composite Motor Car	77	45' 0"	9' 6"	20	50	Dick Kerr	1905	1000
L22	11	Composite Motor Cars	78	45' 0"	9' 6"	20	50	L&YR	1906	1001 - 1011
N22	6	Third Class Motor Cars	73	60' 4¾"	10' 0"	-	(66) 78	L&YR	1906	3034 - 3039
O22	6	First Class Trailer Cars	74	60' 4¾"	10' 0"	64	-	L&YR	1906	428 - 433
P22	2	Third Class Motor Cars	73	60' 4¾"	10' 0"	-	(66) 78	L&YR	1906	3015 & 3023
C23	6	Third Class Trailer Cars	81	60' 4¾"	10' 0"	-	(80) 90	L&YR	1906	3112 - 3117
R23	5	Third Class Motor Cars	88	60' 4¾"	10' 0"	-	80	L&YR	1907	3040 - 3044
S23	2	Third Class Motor Cars	89	60' 4¾"	10' 0"	-	68	L&YR	1907	3045 & 3046
W23	2	Third Class Motor Cars	89	60' 4¾"	10' 0"	-	68	L&YR	1908	3047 & 3048
D24	1	Third Class Motor Car	88	60' 4¾"	10' 0"	-	80	L&YR	1908	3051
V24	6	Third Class Trailer Cars	96	60' 4¾"	10' 0"	-	96	L&YR	1908	3118 - 3123
	2	Third Class Motor Cars	56A	60' 4¾"	10' 0"	-	80	Rebuild	1909	3012 & 3013
	1	Third Class Motor Car	56B	60' 4¾"	10' 0"	-	78	Rebuild	1910	3009
	1	Third Class Control Trailer	68A	60' 4¾"	10' 0"	-	86	Rebuild	1910	3100
	1	Third Class Control Trailer	81A	60' 4¾"	10' 0"	-	89	Rebuild	1910	3117
M26	2	Third Class Motor Cars	89	60' 4¾"	10' 0"	-	68	L&YR	1910	3049 & 3050
P26	1	Third Class Trailer Car	122	63' 7"	10' 0"	-	103	L&YR	1910	3124
D27	2	First Class Trailer Cars	121	63' 7"	10' 0"	76	-	L&YR	1911	434 & 435
E27	2	Third Class Trailer Cars	122	63' 7"	10' 0"	-	103	L&YR	1911	3125 & 3126
A28	4	Third Class Control Trailers	127	63' 7"	10' 0"	-	97	L&YR	1912	3127 - 3130
B28	2	First Class Trailer Cars	121	63' 7"	10' 0"	76	-	L&YR	1912	436 & 437
C28	4	Third Class Motor Cars	128	63' 7"	10' 0"	-	68	L&YR	1912	3052 - 3055
D28	4	Third Class Motor Cars	128	63' 7"	10' 0"	-	68	L&YR	1913	3056 - 3059
M28	14	Third Class Trailer Cars	122	63' 7"	10' 0"	-	103	L&YR	1913	3131 - 3144
D29	6	Third Class Motor Cars	135	63' 7"	10' 0"	-	83	L&YR	1914	3060 - 3065
E29	6	Third Class Trailer Cars	122	63' 7"	10' 0"	-	103	L&YR	1914	3145 - 3150
F29	3	First Class Trailer Cars	121	63' 7"	10' 0"	76	-	L&YR	1914	438 - 440
M31	1	Motor Baggage Car	147	54' 0"	9' 0"	-	-	L&YR	1921	3066

Appendix Three

Carriage Orders completed at Newton Heath between June 1909 and March 1910

O/N	Qty	Type	Diagram	Length	Compartments 1st	Compartments 3rd	Compartments Lav	Compartments Lug	Compartments Gd	Seats 1st	Seats 3rd	Completed	Cost	Notes
D25	3	Third Class	98	54' 0"		9					108	1909	£746-14-10	
E25	12	Third Class	98	54' 0"		9					108	1909	£781-11- 9	
F25	42	Third Brake	94	54' 0"		5			1		60	1909	£718- 7- 9	
G25	6	First Class	99	56' 0"	2					40		1909	£1,478- 4- 2	Centre Corridor & Gangways
H25	1	Composite	100	56' 0"	1	2	1	1		18	32	1909	£1,323- 5- 8	Centre Corridor & Gangways
K25	2	Third Class	101	56' 0"		3	1				66	1909	£1,251- 2- 5	Centre Corridor & Gangways
L25	4	Third Class	55	56' 0"		2	1				66	1909	£1,197- 6- 6	Centre Corridor & Gangways
M25	4	Third Brake	102	54' 0"		2			1		48	1909	£1,029-18- 8	Centre Corridor
N25	3	Compo Brake	84	56' 0"	2	4	2		1	12	28	1909	£1,128- 0- 6	Side Corridor & Gangways
O25	10	Composite	104	56' 0"	4	3	2			24	24	1909	£1,194-16- 1	Side Corridor & Gangways
P25	1	Third Class	103	56' 0"		3	1	1			64	1909	£1,256- 3- 6	Centre Corridor & Gangways
R25	2	Third Brake	94	54' 0"		5			1		60	1909	£701- 6- 8	
S25	21	Third Brake	105	56' 0"		5	1				40	1909-10	£908-18- 8	Side Corridor & Gangways
T25	4	Third Brake	91	56' 0"		4	1				32	1909	?	Side Corridor & Gangways
V25	1	Composite	67	54' 0"	4	4				32	48	1909	£896- 3-10	Oldham Stock
W25	1	Composite	67	54' 0"	4	4				32	48	1909	£907-17- 0	Oldham Stock
Y25	7	Third Brake	66	54' 0"		9					108	1909-10	£772-17-10	Oldham Stock
A26	1	Third Brake	65	54' 0"		6					72	1909	£735-15- 5	Oldham Stock
B26	1	Third Brake	65	54' 0"		6					72	1909	£780-11- 3	Oldham Stock
C26	12	Third Class	97	56' 0"		2	1				64	1909-10	£1,005- 7- 7	Centre Corridor & Gangways
D26	62	Third Brake	94	54' 0"		5	1		1		60	1910	£729- 4- 8	
E26	2	First Class	124	56' 0"	2					40		1910	£1,372-17- 0	Centre Corridor & Gangways
F26	6	First Class	106	56' 0"	8					64		1910	£1,080-16- 9	
G26	5	Third Class	98	54' 0"		9					108	1910	£794-10- 8	

Appendix Four

Wagon types constructed at Newton Heath 1909 to 1910

Type	1909	1910	Notes
Open Goods wagons	192	510	
Coal & Coke Wagons	40	30	20-ton capacity
Half-Box Wagons	378	448	
Covered Goods wagons	710	226	
Cattle Wagons	-	15	
Goods Brake Vans (20-ton)	31	19	
Ballast Wagons	12	12	2 of the 1910 build were of all-steel construction
Sleeper Wagons	12	-	
Rail Wagons	6	-	6-wheeled vehicles
TOTAL	1,371	1,250	

Appendix Five

Locomotive Orders completed at Horwich 1910 to 1921

Lot	Works Nos.	Engine Nos.	Type	Class	Notes
64	1081-1091	1535-1545	2-4-2T	K2	Saturated boilers with Belpaire fireboxes
65	1092-1111	Various	0-4-0ST	B7	
66	1112-1131	Various	0-8-0	Q3	Coal engines with large saturated boilers
67	1132-1151	Various	2-4-2T	K2	Superheated boilers with Belpaire fireboxes
68	1152-1171	Various	0-6-0	F22	Superheated boilers with Belpaire fireboxes
69	1172, 1173	Nos.17&18	0-4-0T		Railmotor engines
70	1174-1193	1546-1565	0-8-0	Q4	Large superheated boilers with Belpaire fireboxes
71	1194-1213	1566-1585	0-8-0	Q4	Large superheated boilers with Belpaire fireboxes
72	1214-1223	Various	0-8-0	Q4	Large superheated boilers with Belpaire fireboxes
73	1224-11243	Various	0-8-0	Q4	Large superheated boilers with Belpaire fireboxes
74	1244-1263	Various and 1586-1598	0-8-0	Q3	Coal engines with large saturated boilers
75	1264-1273	Various and 1599-1603	0-6-0	F19 & F22	
76	1274-1293	1604-1623	0-8-0	Q4	Large superheated boilers with Belpaire fireboxes
77	1294-1313	1624-1643	0-8-0	Q4	Large superheated boilers with Belpaire fireboxes
78	1314-1318	1644-1648	0-8-0	Q4	Large superheated boilers with Belpaire fireboxes
79	Rebuilds	(1506-1525)	4-6-0	N1	4-cylinder superheated passenger engines
80	1319-1328	1649-1658	4-6-0	N1	4-cylinder superheated passenger engines

Appendix Six

Locomotives built at Horwich 1910 to 1921

No.	Cl.	Built	No.	Cl.	Built	No.	Cl.	Built	No.	Cl.	Built	No.	Cl.	Built
1536	K2	3/10	111	K2	3/11	1558	Q4	4/13	673	Q4	7/17	1620	Q4	10/19
1537	K2	3/10	112	K2	3/11	1559	Q4	5/13	694	Q4	8/17	1621	Q4	10/19
1538	K2	3/10	151	K2	4/11	1560	Q4	5/13	697	Q4	8/17	1622	Q4	11/19
1539	K2	3/10	162	K2	4/11	1561	Q4	5/13	698	Q4	9/17	1623	Q4	11/19
1540	K2	3/10	188	K2	4/11	1562	Q4	5/13	789	Q4	9/17	1624	Q4	12/19
1541	K2	3/10	221	K2	5/11	1563	Q4	6/13	915	Q4	9/17	1625	Q4	12/19
1542	K2	4/10	223	K2	5/11	1564	Q4	6/13	919	F19	7/17	1626	Q4	1/20
1543	K2	4/10	224	K2	5/11	1565	Q4	6/13	921	F19	8/17	1627	Q4	1/20
1544	K2	4/10	227	K2	5/11	1566	Q4	7/13	923	F19	9/17	1628	Q4	3/20
1545	K2	4/10	230	K2	6/11	1567	Q4	7/13	925	F19	10/17	1629	Q4	3/20
2	B7	4/10	253	K2	6/11	1568	Q4	8/13	926	F19	10/17	1630	Q4	4/20
3	B7	4/10	275	K2	6/11	1569	Q4	8/13	927	Q3	12/17	1631	Q4	5/20
8	B7	5/10	276	K2	7/11	1570	Q4	9/13	1351	Q3	12/17	1632	Q4	5/20
12	B7	5/10	285	K2	7/11	1571	Q4	9/13	1354	Q3	12/17	1633	Q4	5/20
17	B7	5/10	480	K2	7/11	1572	Q4	9/13	1357	Q3	1/18	1634	Q4	6/20
19	B7	5/10	519	K2	10/11	1573	Q4	10/13	1365	Q3	1/18	1635	Q4	6/20
28	B7	5/10	618	K2	10/11	1574	Q4	10/13	1368	Q3	2/18	1636	Q4	7/20
43	B7	5/10	637	K2	11/11	1575	Q4	10/13	1369	Q3	2/18	1637	Q4	7/20
56	B7	6/10	No.17		12/11	1576	Q4	11/13	1586	Q3	4/18	1638	Q4	8/20
64	B7	6/10	No.18		12/11	1577	Q4	11/13	1587	Q3	4/18	1639	Q4	8/20
71	B7	6/10	657	F22	3/12	1578	Q4	11/13	1588	Q3	5/18	1640	Q4	8/20
75	B7	6/10	691	F22	3/12	1579	Q4	12/13	1589	Q3	6/18	1641	Q4	9/20
118	B7	6/10	791	F22	3/12	1580	Q4	12/13	1590	Q3	7/18	1642	Q4	9/20
226	B7	6/10	883	F22	4/12	1581	Q4	12/13	1591	Q3	8/18	1643	Q4	10/20
271	B7	6/10	907	F22	4/12	1582	Q4	12/13	1592	Q3	8/18	1644	Q4	10/20
298	B7	6/10	917	F22	4/12	1583	Q4	1/14	1593	Q3	9/18	1645	Q4	11/20
481	B7	6/10	918	F22	4/12	1584	Q4	2/14	1594	Q3	9/18	1646	Q4	12/20
517	B7	6/10	1363	F22	5/12	1585	Q4	2/14	1595	Q3	10/18	1647	Q4	12/20
613	B7	7/10	1364	F22	5/12	6	Q4	2/14	1596	Q3	11/18	1648	Q4	1/21
614	B7	7/10	1366	F22	5/12	7	Q4	3/14	1597	Q3	11/18	1522	N1	11/20
9	Q3	7/10	209	F22	5/12	134	Q4	4/14	1598	Q3	12/18	1514	N1	1/21
29	Q3	8/10	234	F22	6/12	141	Q4	5/14	1599	F22	7/18	1523	N1	1/21
35	Q3	8/10	239	F22	6/12	142	Q4	6/14	1600	F19	9/18	1510	N1	2/21
67	Q3	8/10	300	F22	6/12	150	Q4	7/14	1601	F22	9/18	1511	N1	2/21
87	Q3	9/10	20	F22	6/12	229	Q4	8/14	1602	F19	10/18	1506	N1	3/21
96	Q3	9/10	219	F22	7/12	259	Q4	9/14	1603	F19	10/18	1521	N1	3/21
616	Q3	10/10	233	F22	7/12	281	Q4	10/14	1604	Q4	11/18	1516	N1	3/21
617	Q3	10/10	237	F22	8/12	460	Q4	12/14	1605	Q4	12/18	1525	N1	3/21
628	Q3	10/10	243	F22	9/12	216	Q4	9/16	1606	Q4	2/19	1509	N1	4/21
713	Q3	10/10	920	F22	10/12	235	Q4	9/16	1607	Q4	2/19	1520	N1	4/21
902	Q3	11/10	1546	Q4	11/12	241	Q4	10/16	1608	Q4	2/19	1517	N1	5/21
905	Q3	11/10	1547	Q4	12/12	265	Q4	11/16	1609	Q4	3/19	1524	N1	6/21
906	Q3	11/10	1548	Q4	12/12	312	Q4	11/16	1610	Q4	3/19	1518	N1	6/21
908	Q3	12/10	1549	Q4	1/13	493	Q4	12/16	1611	Q4	3/19	1519	N1	6/21
909	Q3	12/10	1550	Q4	1/13	503	Q4	12/16	1612	Q4	4/19	1649	N1	8/21
910	Q3	12/10	1551	Q4	1/13	601	Q4	3/17	1613	Q4	6/19	1650	N1	8/21
911	Q3	1/11	1552	Q4	2/13	603	Q4	4/17	1614	Q4	6/19	1651	N1	8/21
912	Q3	1/11	1553	Q4	2/13	507	Q4	4/17	1615	Q4	6/19	1652	N1	8/21
913	Q3	2/11	1554	Q4	2/13	626	Q4	5/17	1616	Q4	7/19	1653	N1	9/21
914	Q3	2/11	1555	Q4	3/13	631	Q4	5/17	1617	Q4	7/19	1654	N1	10/21
18	K2	3/11	1556	Q4	3/13	660	Q4	6/17	1618	Q4	8/19	1655	N1	9/21
70	K2	3/11	1557	Q4	4/13	640	Q4	7/17	1619	Q4	8/19	1656	N1	11/21

The last two locomotives of Lot 80, Nos. 1657 & 1658, were completed in January 1922

Appendix Seven

Progress of National Projectile Factories for two weeks ending 15th April 1916

National Projectile Factory	Total m/cs including Tool Room	Received 15th April	Received 1st April	Received Increase	MACHINES Fixed 15th April	Fixed 1st April	Fixed Increase	Working 15th April	Working 1st April	Working Increase	Est Total Ultimate Staff	No. of Employees on Production 15th April	1st April	Increase	No. of Shells in Progress 15th April	1st April	Increase	Total Production to April 15th Ø
Harper Bean & Son	1,004	545	467	78	310	268	42	147	91	56	3,450	346	246	100	3,217	1,684	1,533	60-Pdr Shrap = 15
Dick Kerr & Co	1,194	541	463	78	521	463	58	269	228	41	3,000	454	324	130	5,648	3,952	1,696	6" H.E. = 200
T. Firth & Sons Ltd	1,389	884	771	113	656	550	106	319	300	19	5,300	880	382	498	12,612	2,276	10,336	60-Pdr H.E. = 243
Beardmore & Co	1,138	596	547	49	321	250	71	193	179	14	3,100	385	267	118	No Return	5,733	N/A	
Vickers Ltd	1,610	833	833	0	567	499	68	114	76	38	5,360	296	169	127	8,537	6,212	2,325	60-Pdr Shrap = 233
Babcock & Wilcox	468	316	301	15	60	40	20	0	0	0	1,135	0	0	0	0	0	0	
Leeds NPF	175	97	64	33	72	44	28	39	29	10	2,000	50	28	22	103	61	42	
Cammell Laird & Co	805	456	375	81	365	317	48	50	40	10	3,050	1,073	620	453	1,644	507	1,137	
Hadfields Ltd	589	439	379	60	382	276	106	125	49	76	2,200	153	56	97	410	233	177	
Armstrong Whitworth & Co	1,063	412	165	247	267	140	127	0	0	0	0	0	0	0	0	0	0	
G&J Weir Ltd	262	131	131	0	73	60	13	14	14	0	545	52	37	15	486	364	122	
Totals	9,697	5,250	4,496	754	3,594	2,907	687	1,270	1,006	264	29,140	3,689	2,129	†1,560	32,657	15,289	17,368	
		54.1%	46.4%	7.8%	37.1%	30.0%	7.1%	13.1%	10.4%	2.7%		12.7%	7.3%	5.4%				

Ø Shells practically complete
‡ Beardmore omitted from the total

Appendix Eight

Summary of Up Trains, Deir Sineid – Junction Station
21st November to 7th December 1917

Date	Trains	Axles	Tons	
21 Nov 1917	1st	40	120	
22 Nov 1917	Nil	-	0	
23 Nov 1917	1st	40	125	
24 Nov 1917	1st	36	115	
	2nd	40	170	
25 Nov 1917	1st	36	130	
26 Nov 1917	1st	16	100	
	2nd	40	200	
27 Nov 1917	1st	30	176	
	2nd	30	160	
28 Nov 1917	1st	34	224	
	2nd	16	80	
	3rd	22	80	R.E. Construction Coy.
	4th	20	120	
29 Nov 1917	1st	8	32	H.Q. Staff Corps
	2nd	34	136	
30 Nov 1917	1st	32	90	
	2nd	24	96	
	3rd	26	104	
	4th	28	112	
1 Dec 1917	1st	12	48	
	2nd	18	72	
	3rd	38	128	
	4th	30	120	
2 Dec 1917	1st	98	368	
	2nd	18	72	
	3rd	16	64	
	4th	32	128	
	5th	18	64	
	6th	12	48	
	7th	34	100	Ammunition Train
3 Dec 1917	1st	130	476	
	2nd	14	56	
	3rd	32	120	
	4th	38	136	
	5th	16	64	
	6th	18	72	
	7th	34	104	
	8th	28	100	Ammunition Train
4 Dec 1917	1st	16	40	
	2nd	16	64	
	3rd	18	72	
	4th	38	112	
	5th	14	56	
	6th	18	48	
	7th	24	90	Ammunition Train
5 Dec 1917	1st	32	104	
	2nd	16	64	
	3rd	36	144	
	4th	16	56	
6 Dec 1917	1st	16	64	
	2nd	40	104	
7 Dec 1917	1st	16	64	
	2nd	40	160	
	3rd	16	64	
	4th	34	100	Ammunition Train

From Major C.E. Jordan-Bell's report of 8th December 1917 to Col. H.E. O'Brien

Appendix Nine

Electric Traction

By Col. H.E. O'Brien, D.S.O., M.I.C.E., M.I.E.E.

Paper read before the Irish Centre of the Institution of Electrical Engineers, Dublin

15th January 1931

To the passenger sitting in a more or less comfortable carriage, or to the trader who wishes his goods conveyed to one spot from another, it may seem to make little difference as to whether a train is hauled by electric or steam power. Yet, if to both these classes is offered more rapid transit from place to place, and to the former much greater cleanliness both on the inside and outside of his compartment, and if electric haulage offers these advantages, then the general public, to a lesser degree indeed than the railway companies, should be interested in electric traction. To the railway companies any change of method of haulage likely to enable them to offer lower rates and fares, more comfort to their customers and higher dividends to their shareholders must indeed be of absorbing interest.

There is no question as to the greater comfort, higher average speed and lower working expenses obtainable on railways with heavy grades by changing from steam to electric traction. Most of the railway companies in the world operating over heavy grades have either electrified, or are in the process of electrifying, or will electrify as soon as power is available. The advisability of making such a change on railways operating on lines with grades not greater than 0.5% to 1.0% is much more open to question.

Before proceeding to the more detailed consideration of the problems involved in main line electrification, a brief reference to suburban and branch line electrification is desirable.

The cost of equipping and operating an electric suburban service working within a radius of 20 to 30 miles from a large city can be accurately estimated. Whether the cost of operating such an electric service is going to be greater or less than that of the existing steam service depends on several factors. If the scheduled speed is going to be increased, and also the frequency of the service, it is probable that the cost per train mile will not be increased, but the total cost of the service will be increased. Experience of many such electrifications already accomplished indicates that these electrifications have been made with the expectation of obtaining large increases of traffic, an expectation which has invariably been fulfilled. These increases are generally as much due to the greater frequency of service, which can be effected much more easily with electric than with steam haulage, as to the increase in scheduled speed which, however, has its attractions for the public.

It may be as well to just recapitulate the advantages of suburban electrification. The movements at the terminals are reduced as on arrival at the platform the driver and guard have only to change places and the train is ready to start again, hence the amount of terminal accommodation required is much reduced and also the signalling movements. At Liverpool three platforms suffice to deal with a five-minute service in the rush hour. The maintenance of the rolling stock costs much less than that of steam locomotives performing an equivalent service. The trains are operated by one man instead of two. The trains can stay in service for long periods, as there are no locomotive duties such as coaling, watering, fire cleaning, etc. Running maintenance is greatly simplified as with well-designed rolling stock inspection once a week, or at the most twice, is sufficient; this inspection in general can be carried out anywhere, in sidings or at platforms.

In general, however, it may be said that suburban electrification is not worth carrying out unless there is a considerable density of traffic and definite prospect of further developments and also a supply of cheap current available. Any train service with a frequency of two trains per hour each way is worth investigating, but each case must be dealt with on its merits. So much work has been accomplished in this sphere of electrification, and so much written about the technical and other sides of the subject, that I do not propose to deal further with it tonight.

Branch lines usually have a sparse traffic and are of inconsiderable length: they do not lend themselves to electrification except at a loss as when unavoidably embraced in a general scheme of electrification.

In considering the possibilities of the successful electrification of a main line devoid of heavy grades so many variable factors have to be considered that it is impossible to deal with the subject in any adequate manner in the course of a short lecture.

As a premise it may be stated that whereas the overall efficiency of the steam locomotive of a railway from coal pile to tread of wheel is from 6% to 8%, the efficiency of a system using electric locomotives is from 12% to 16% or, put more simply, if one pound of coal contains 100 units of energy a steam locomotive can put 6 to 8 of these into hauling the train and the electric locomotive 12 to 16; the rest of the energy is wasted. But actually, a cheaper grade of coal can be burnt in generating electricity for a steam locomotive requires a higher grade of coal.

Therefore, as the electric system only requires half as much coal as the steam system, it is clear that the cost of supplying energy to the trains will be the same in both cases if approximately half the cost of electrical energy is the cost of coal burnt at the generating station. An actual example will help: assume that 2,000 lb. of coal costing 30s 0d will enable a steam locomotive to haul a train 40 miles. If the train is hauled electrically 1,000 lb. of coal costing 10s 0d will provide the necessary units, the remaining 20s 0d is available to meet interest on capital, depreciation, operating costs, etc. for all the plant necessary to produce the electric current and convey it to the trains, viz. generating station, high tension transmission lines, converter or transformer stations, third rail or overhead trolley.

In connection with the above it must be premised that the electric generating station is of the most efficient modern type operating at a load factor of not less than 40% with an overall efficiency of 20% to 25%. If the station is a hydro-electric one the position is essentially the same only the interest, depreciation and maintenance charges on the cost of works for leading the water to the turbines is taken instead of the cost of coal.

We thus come naturally to the question of load factor. The load factor of a railway system may be very high, one in fact to make the mouth of a supply engineer water. Take the elementary case of a single trunk line 120 miles long carrying a traffic of freight and passenger trains moving from one terminal to another at a uniform speed of 40 mph. The trip from one end of the line to the other will take a train 3 hours; a double trip takes 6 hours. Four double trips per day with only one train in occupation of the line will give 100% load factor. This is an extreme and elementary case, but it is worth developing a little further. Let us assume that the single line is equipped with a single-phase 16,000-volt trolley, that the line equipment costs £120,000 and that the annual charge for interest, depreciation and maintenance on this is £12,000 per annum. The daily train mileage will be 960 and the annual train mileage 350,000. Thus the charges per train mile, in respect of the line equipment only, will be nearly 8 pence per train mile. If trains of 400 tons (including the weight of the locomotive) are used, the electrical units consumed per train mile at 30 watt-hours per ton-mile will be 12kW-h. If electrical energy can be purchased delivered to the trolley at ½d per unit, the cost of energy will be 6d and the total cost of energy delivered at the locomotive 6d + 8d = 14d per train mile.

For steam operation of this service the coal consumption would probably be about 40 lb. per mile, and if locomotive coal costs $^1/_{10}$d per lb. = 18s 8d per ton, the fuel cost is 4d per mile. But, if locomotive coal costs 37s 4d per ton on the one hand, and on the other hand the traffic is trebled and increased to twelve double trips per day, the cost of the electrical energy at the locomotive per train mile will become 6d + $^8/_3$d = $8^2/_3$d while the fuel cost with steam operation will become 8d. This shows, in an elementary way, the effect of variations in the cost of coal and in the density of traffic on the cost of energy at the locomotive.

Unfortunately, no main line approaches the ideal conditions of the simple illustration we have just considered. Passengers want to travel at certain hours of the day and not at others; freight must be dispatched at night in order to arrive at its destination the following morning; the track is generally double and includes possibly junctions and many sidings thus greatly increasing the cost of the track equipment; and the electric power is not always purchasable at as low a cost as ½d per unit. Still, on a principal artery of traffic joining two important commercial centres it will be found that the traffic density is not far from constant and that it is of such a density as to make the charge for electric energy per train mile not much in excess of the cost of locomotive coal. I have shown you two slides representing a graphical timetable for a typical busy double-track mainline and the consequent load which would be produced if electric working was resorted to.

It may be here remarked that, if current is purchased, as a matter of strict accountancy the only true capital charge incurred in electrifying a main line railway is that for the equipment of the track; that is on the basis of the same service and same schedule speed being maintained as with steam locomotive operation. If, however, the electric locomotives provided are to have a greater haulage power than the steam locomotives then clearly a portion of the expenditure on the new locomotives is chargeable to capital and not to renewals.

The cost of the electric equipment of the track of a mainline varies very much; taking at the one end of the scale a single or double-track line with few junctions, sidings and tunnels, and operating in a pure atmosphere with a comparatively infrequent service, the cost might be as low as £1,000 per mile using 16,000 volts single phase. At the other end of the scale a busy main line with four or more trains per hour each way, with many junctions, sidings, tunnels and over-bridges, might cost anything up to £4,000 per mile on either the 1,500-volt direct current or on the high voltage single-phase system.

Let us now consider the question of the other expenses involved in the operation of locomotives. We have dealt with the cost of fuel; what about the cost of repairs, the wages and other expenditure at running sheds, and the wages of driver and fireman? Before considering these, it will be well to deal with the design of the electric locomotive. The electric locomotive has a number of great advantages as compared with the steam locomotive: it may be much lighter in weight for the same tractive effort; it has no reciprocating parts causing racking and twisting stress on the frames; it can be safely operated by one man; it exerts an even and constant effort on the driving wheel treads; it makes no dirt; and no parts work at high temperatures.

The most usual design of electric locomotive is exactly similar in general principles to a tramcar; the superstructure containing the controller, switches and resistances may or may not also combine a passenger compartment. This superstructure is carried on two bogies, each of which is usually equipped with two motors. Such locomotives for 4ft 8½in. gauge may be rated up to 1,200 hp for one hour and can exert a tractive force equal to that of the most powerful steam locomotive. This type of locomotive is extensively used for medium speed passenger trains and freight trains, large numbers being in use in France, the United States and elsewhere. Owing to the weight of the motors on the bogie and the consequent low centre of gravity, this type causes heavy wear on the rails on curves and is also liable to bogie oscillation at high speeds. However, by using a long wheelbase bogie and as light motors as practicable it is possible to run this type up to 70 mph on lines without severe curves. It is the type universally used for suburban trains. The cost of maintenance of this type of the most modern design is well established; it should not exceed 3 pence per train mile.

The problem of high speeds and also the desire to reduce the number of motors to a minimum has produced other types, several of which have proved most successful in practice. It may be accepted as an axiom that a reduction of the number of parts in any mechanism means that (a) the cost of running maintenance and inspection is reduced, and (b) if the design and construction is sound, the cost of periodical overhaul is also reduced. For instance, it takes less time to examine the brush gear and replenish the oil supply of two 500hp motors than of four 250hp motors; a gear drive in an oil bath requires less attention than a set of coupling rods; the switchgear for two motors is less expensive and less complicated than for four or six motors.

These designs fall into several distinct classifications:

(1) Gearless — armature direct on axle, field coils on the frame;

(2) One or two large motors placed above the level of the wheels and driving theses by connecting rods and coupling rods;

(3) Motors placed above the level of the wheels with either vertical or horizontal shafts driving a hollow quill or sleeve, this sleeve surrounds the axle which it drives by means of a flexible linkage;

(4) Motors placed horizontally above the level of the wheels driving a gear wheel outside and in the same plane as the wheels, these gear wheels are connected to the driving wheels by a flexible linkage.

Type No.1 is the simplest and most robust of all the designs, no gears, no coupling rods. It has the disadvantage of a low centre of gravity, a considerable weight on each axle not spring borne, and a limited power on each axle. The Paris Orleans Railway has a notable example of this type in operation.

Type No.2 has been and is much used on the Continent. It has the disadvantages of employing coupling rods, which have to be kept in accurate adjustment; and of converting rotary into reciprocating and back into rotary motion. On the other hand, it has the advantages that one or two large motors can be employed to drive a number of driving wheels and that the wheels and axles are easily removable for inspection. This type has a future for slow freight traffic.

Type No.3 has two disadvantages — the linkages between the sleeves and axles prove very difficult to lubricate and require frequent renewal and the axle cannot be inspected without the removal of the sleeve.

Type No.4 has many advantages. The motors are easily accessible; the linkage between the gear wheels and driving wheels is of a simple character and is easily lubricated; and it is a very simple matter to withdraw the wheels and axles which can then be inspected in their entirety free from any appendages such as quills.

The duty imposed on the whole equipment of a main line electric locomotive is much less severe than that imposed on the electrical and running equipment of suburban electric trains. If it is assumed that the annual mileage of the electric tractor is 60,000 miles per annum, the probability is that shocks of starting, accelerating and braking will be imposed once every mile on the suburban equipments and probably not more than once every 20 miles on the main line equipments. It is obvious that this means that the cost of repairs, running maintenance and inspection should be much smaller for the main line locomotive than for the suburban motor car. From my experience of the maintenance of suburban equipments, and from what I have seen of the most recent main line locomotive designs on the Continent, I have no doubt that in the near future the main line electric locomotive will only require to come out of traffic for wheel turning and tyre replacement. The workshops of the future on an electrified main line will consist of a small armature and field coil rewinding shop, a shop for the periodic overhaul of contactors and switchgear, and a wheel and tyre shop.

The locomotives themselves will remain continuously in service except for brief visits to the running sheds at comparatively long intervals to have a set of wheels or contactors changed. On the L&NER, Sir Vincent Raven has stated that on freight work this cost of repairs and shed wages was $1/7$ of that of steam locomotives doing exactly similar work. The complete elimination of all water, coal and ash services and the time spent in engine duties, which is very heavy for some steam services, is also completely eliminated.

One of the traffic difficulties in main line electrification, and indeed also in suburban electrification, is the proper utilisation of the long hours of continuous service available from the electric locomotive. Take an illustration on an actual test over a 90-mile run with one of the most modern and most powerful four-cylinder steam locomotives capable of developing a maximum horsepower at the rails of 1,250 and hauling a heavy train — both smokebox and fire required cleaning and fresh supplies of coal on the tender were also required. An electric locomotive could not only have recommenced the return journey at once, but could also have continued making 90-mile trips for at least a week without attention to any detail except lubrication. On the Chicago, Milwaukee & St. Paul Railroad a single locomotive has now run 766 miles in 24 hours and a continuous run of 440 miles is regularly made.

At present, for some not very clear reason, an electric locomotive costs 60% more to purchase than a steam locomotive of equal power. Therefore, to equalise the standing charges the electric locomotive must run 60% more mileage per annum. This is a problem for the traffic department to arrange the workings of the locomotives so as to give the electric locomotive that continuous service of which it is capable.

The use of electric locomotives permits of a considerable increase in the average speed of the train services, particularly on lines with gradients, owing to the possibility of maintaining the speed on rising gradients and also due to the elimination of stops for engine duties. The Midi and Paris Orleans railways in France have found that the schedule times for many of their trains can be reduced by 10% to 20%.

There is a good deal of misconception about the relative cost of steam and electric services due to comparisons not being made on an equal basis. For instance, the Swiss Federal Railways are believed not to have reduced their operating costs appreciably, but it is forgotten that their system is one with rather sparse traffic, with many junctions and tunnels, that a great deal of the construction was carried out at war prices, that a great deal of money was spent on experimental locomotives, and that the cost of the energy from the hydro-electric stations is rather high owing to the immense amount of storage that had to be provided to meet the winter annual water deficiency. On the other hand, it must not be forgotten that a faster and more frequent service of trains is being provided.

A reduction in the time schedule of the trains and the elimination of engine duties automatically reduces the cost of wages per train mile for the train crews as these are paid by time and not by mileage. But a further and very substantial economy in operation arises from the possibility of operating the locomotive with only one man instead of two. In Switzerland one-man operation is already in force on the Rhaetian Railway and on some of the locomotives of the Lochtschberg Railway. On the Federal railways the number of drivers in 1913, when the working day was 10½ hours, was 3,250. Now, with an 8-hour day, it is only 3,172. *The Times* of 29th October 1926 stated that use of one-man driven electric locomotives was to become general in spite of the strong opposition of the trade unions.

Another point in favour of the electric locomotive is its lower total weight as compared with the steam locomotive and the ease with which the weight per foot can be reduced if required. On a test made with two of the most powerful modern four-cylinder engines hauling a 500-ton passenger train, and later a 640-ton freight train, the maximum drawbar pull recorded was under 15 tons. This means that 60 tons adhesive weight on the driving wheels would have sufficed to meet these maximum conditions but, actually, the two steam locomotives necessary to maintain schedule with these heavy test trains aggregated 237 tons. There were 6 driving axles, each carrying 20 tons. An electric locomotive with four driving axles each carrying 18 tons, and with an aggregate weight of not more than 112 tons would have been capable of hauling these heavy trains at the same speed as the steam locomotives.

I recently travelled on one of the Paris Orleans Railway locomotives, which had a maximum axle load of 12.6 tons, a weight per foot run of length over buffers of 1.73 tons, and capable of exerting 2,550hp at the drawbar at a speed of 56 mph. This locomotive hauled a 500-ton passenger train at 80 mph on a very bad track.

It will be seen, therefore, that the smaller weight of the electric locomotive permits of the postponement of bridge renewals and results in reduced wear, not only on the track but also of its own tyres. I will now recapitulate the principal points to be borne in mind in connection with electric traction schemes:

1. The only true capital expenditure is on the equipment of the track, for locomotives

should be provided on renewal account — this is only true if the train services and schedules are unaltered, if they are increased a proportion of the cost of the locomotives is debitable to capital. Current should be purchased from outside sources.

2. The prospect of the financial success of any electrification is dependent to a considerable extent on the relationship between the cost of current on the one hand, and the cost of coal on the other hand, at the locomotive. This relationship depends to a very considerable extent on the density of traffic on the lines electrified because the fixed charges on the track equipment are spread over a great many units of electricity in the case of dense traffic and over only a few in the case of light traffic

3. Given that the cost of coal and current are the same, the electric locomotive may be expected to cost less than half than does the steam locomotive for repairs and maintenance, and about half for wages of traincrews if one-man operation is adopted.

4. The adoption of electric traction offers opportunities for improving railway services, which do not offer themselves with steam traction, and may affect important economies in operation.

There are a number of advantages of electric traction which cannot be given exact monetary values. Tunnel repairs will be reduced, especially for tunnels on grades where a heavy blast from the steam locomotive exhaust strikes the roof of the tunnel. A certain amount of sidings now used for coal storage and ash removal will be freed for other purposes. The whole electrified line becomes much cleaner owning to the absence of smoke and ashes, and the passenger carriages also remain much cleaner as they are no longer struck by the small gritty cinders from the steam locomotive. Fire risks are reduced.

Having dealt with electric traction from a general stand point, it is now necessary to say a few words about the choice of systems. At the present moment 1,542 miles in Switzerland have been electrified on the single-phase system at 16,000 volts; 1,013 miles in Italy on the three-phase system at 3,000 volts; Germany has adopted the single-phase system while Holland and France have standardised direct current at 1,500 volts for many miles of track already and about to be electrified. America too appears inclined to standardise 1,500 volts direct current as is also Great Britain.

There seems to be little doubt that 1,500 volts direct current is the best system to adopt for a suburban electrification and for mainlines with heavy traffic; the standard frequency of 50 cycles can be used for the main supply thus enabling the same H.T. transmission system to feed both the railways and the general supply of the country. The locomotives are somewhat cheaper and simpler to build and maintain, and difficulties with inductive and capacity effects in the telephone and telegraph systems are less. The single-phase system at 16,000 volts is, however, particularly suitable for long lengths of mainline with rather sparse traffic; the three-phase system is not likely to have any extensive use outside Italy as the overhead work is too complicated.

The 1,500-volt system can be operated either from an overhead trolley or from a third rail; the overhead work is rather heavy and expensive and difficult to maintain as it is up in the air. There are two objections to the third rail, firstly the danger it offers to staff and passengers, and secondly the difficulty of maintaining continuity of contact for locomotives (not for multiple-unit trains) at junctions. The danger to passengers and staff can be largely eliminated by the use of an under-contact third rail with a special moulded protection (such a rail costs about £2,000 per mile erected but exclusive of special work). The second objection is not serious.

The 16,000 volts single-phase system offers the cheapest overhead construction as the trolley wire is very light, but the desirability of using the low frequency of $16^2/_3$ cycles necessitates the use of frequency changers if the general supply of the country is to be made use of.

No doubt you will ask me whether there is any prospect of any of the Irish railways being electrified. The answer is that it is impossible to tell whether a railway can be electrified and the electrification can become a financial success without making an investigation. It is so easy, and so little expense is involved making a preliminary investigation of such a problem which will definitely show whether it is worth going to the expense of making complete plans and estimates, that it is surprising that more railway companies do not go into the matter. Full ton-mileage statistics are very desirable for use in making complete estimates of the prime costs and working costs of an electrified railway, but are not absolutely necessary.

Appendix Ten

An Alpine Adventure

Article written by Col. H.E. O'Brien, 1935

The recent adventure of a man who was trapped between rocks in a cave in England reminded me of an incident 21 years ago on a well-known Alpine peak. The Grepon, one of the precipitous rock needles above Chamonix, is a little over 11,000ft. high. Baedeker describes it as the most difficult of the Chamonix aiguilles; but this is not exactly true, as far more difficult climbs have been discovered since the great climber A.F. Mummery made his first ascent in 1881. An Alpine joke describes this peak successively as an inaccessible mountain, the most difficult climb in the Alps, and easy day for a lady. Well it is certainly not an easy climb; access is made possible by the cracks and fractures in the almost vertical granite precipices. Just on 60 I feel that this climb must be done before I am too old. There is the advantage that the night before the climb can be spent in one's comfortable hotel room. A guide and a porter accompanied me.

Various risks and difficulties were overcome; a hurried passage under frowning and unstable ice cliffs, the 70ft. Knubel crack which has almost no holds, the curious hole which has to be squeezed through. At last we are nearly on top climbing the last crack. The guide is on the small flat summit; my hands are on the edge of the summit. Unfortunately, to help the last pull I put my knee in the crack. I heave, but my knee is stuck fast in the crack. The guide pulls the rope from above; the porter shoves from below, but not a move from my knee. Not only am I stuck fast but also am blocking the way to the top for other parties including two ladies who are following us. What is to be done? Get a surgeon or a stonemason to free my knee? Dilemmas make for quick thinking. Perhaps it is the cloth of my trouser that is holding me. I have a sharp knife in my pocket and slit open the knee of the trouser and out comes my knee. The little adventure is over; the climber slightly shaken. The first part of the descent is a little sensational. The guide disappears into the void and calls me on. I pass the rope under my thigh and over the back of my neck and step off the edge. Looking down all I can see is the glacier 3,000ft. below. I slide down about 40ft. and suddenly find myself on a little rock platform beside the guide; the rest of the descent is comparatively easy.

Some years later I am crossing a level glacier plateau on my way to Italy. A guide leads, next on the rope my nephew, then myself, then a porter. Suddenly the snow gives way. I fall about 10ft. before the rope holds me. First thought, will the rope hold? Swiss guide ropes are tested usually, but not French. It holds; I look up and see the sky through the hole above, down and see two blue walls of ice receding into darkness a hundred feet below. Fortunately I have kept hold of my ice axe and have fallen where the crevasse narrows; though constricted by the rope round my chest, I manage to chip a foothold in the ice wall. Intense activity above; I shout directions, the others pull and out I come into the sunshine. Another little adventure is over.

Appendix Eleven

Fishing Recollections

By Col. H.E. O'Brien

I caught my first salmon in the Dee in 1913 at Banchory but rare the opportunities of a day thereon; ever since for me there has been no river to compare with that lovely stream. In June 1950 I had the good fortune to have a fortnight's fishing on some five miles round but mainly above Potarch Bridge. The river was very low, not difficult to wade across in the shallows, and crystal clear, the water temperature about 60°F; the nights cool, the days warm with bright sun. Every pool was full of fish. One pool had a shingle shore the water gradually deepening to about 4 to 5ft. to a channel bordered by rocky ledges. Dozens of fish could be seen lying on the shingle and in the channel but mainly at the back end of the pool where fast water ran down a rapid. As one waded out, as soon as one got within about 12ft. of the fish they gently moved away towards the far side.

One could watch the prawn or fly move over the fish as clearly as if there had been no water there, but as the fly moved down into the rough water at the lip of the rapid only the greased line indicated its position. The comparatively few taking fish in all pools seemed to lie in the rough water just above the rapid or fall, and in the case of a taking a fish the line was therefore straight down stream, the position of the fish such that it was very difficult to get level with him and prevent him going down the shallow rapid. These conditions combined with the rough wading among the many boulders, which here cover the bed of the Dee, made an exciting single-handed job of playing and gaffing the fish. In spite of the low warm water fresh salmon and sea trout seemed to move up every day. For some days there was a very big fish lying in the very middle of the river a few yards above the small Potarch fall, but he moved up and was seen no more.

From many fishermen's point of view the fortnight was poor fishing; the bag three salmon, 5½, 7, and 8lbs, two sea trout 3 and 2lbs. I lost about the same number and touched about as many again. The year before in September I had a day near Ballater, again in lovely weather, scenery, very low water, pools stiff with fish (rather red and ugly), but for me no takers.

It was interesting to see that the many fish in the pools at Potarch took no notice of the prawn; the same thing is noticeable at Galway. If a prawn really panics fish it surely ought to be possible to make the hundreds of fish at Galway stampede. It is evident that the scaring of fish by a prawn is a rare phenomenon, no fisherman I have met has ever seen it happen. There are similar fables about the otter. Believe it or not a river owner and experienced fisherman was going to have an otter shot because she thought it disturbed her favourite pool. She never seemed to think of the otter's diet of eels and the many salmon parr whose lives it saved. How many of us have ever seen a lamprey? I saw one once on the Shannon, a disgusting looking creature.

Bibliography

Aspinall, J.A.F.: *The Horwich Locomotive Works of the Lancashire & Yorkshire Railway*, Paper No.3009, Proc. I.C.E., Vol.CXXIX, 1896-97

Bachellery, A.: *Electrification of the French Midi Railway*, I.E.E. Journal, Vol.62, No.327, 1924

Barnes, R.: *Locomotives That Never Were*, Jane's Publishing, London, 1985

Blakemore, M.: *The Lancashire and Yorkshire Railway*, Ian Allan, London, 1984

Bond, R.C.: *A Lifetime with Locomotives*, Goose & Son, Cambridge, 1975

Bulleid, H.A.V.: *The Aspinall Era*, Ian Allan, London, 1967

Burke, Sir B.: *Burke's Peerage & Baronetage*, London, various editions

Burke, Sir B.: *Burke's Landed Gentry*, London, various editions

Byrne, Prof. F.J.: *Irish Kings and High Kings*, (2nd edition), Four Courts Press, Dublin, 2001

Carter, F.W.: *The mechanics of Electric Train movement*, I.E.E. Journal, Vol.50, No.218, 1913

Chacksfield, J.E.: *Richard Maunsell – An Engineering Biography*, Oakwood Press, Usk, 1998

Chacksfield, J.E.: *Sir Henry Fowler – A Versatile Life*, Oakwood Press, Usk, 2000

Cleveland-Stevens, E.: *English Railways — Their Development and Relation to the State*, George Routledge & Sons, London, 1915

Coates, N.: *Lancashire & Yorkshire Wagons — Vol.1*, Wild Swan Publications, Didcot, 1990

Collard, W.: *Proposed London and Paris Railway*, Printed for private circulation, 1928

Cook, A.F.: *LMS Locomotive Design and Construction*, RCTS, Lincoln, 1990

Cotterell, P.: *The Railways of Palestine and Israel*, Tourret Publishing, Abingdon, 1984

Cox, E.S.: *Locomotive Panorama — Volume 1*, Ian Allan, London, 1965

Cox, E.S.: *Chronicles of Steam*, Ian Allan, London, 1967

Dalziel, J. & Sayers, J: *The Single-phase Electrification of the Heysham, Morecambe and Lancaster Branch of the Midland Railway*, Paper No.3847, Proc. I.C.E., Vol.CLXXIX, 1909

Davies, W.J.K.: *Light Railways of the First World War*, David & Charles, Newton Abbot, 1967

Davis, R.: *Revolutionary Imperialist – William Smith O'Brien*, Lilliput Press, Dublin, 1998

Dawson, P.: *The Electrification of a portion of the Suburban System of the London Brighton & South Coast Railway*, Paper No.3929, Proc. I.C.E., Vol.CLXXXVI, 1911

Dawson, Sir Philip: *Electric Railway Contact Systems*, I.E.E. Journal, Vol.58, No.295, 1920

Ellis, C. Hamilton: *The Midland Railway*, Ian Allan, London, 1953

Firth, H.W.: *Electrification of Railways as affected by traffic considerations*, I.E.E. Journal, Vol.52, No.234, 1914

Forbes, N.N., Felton, B.J. & Rush, R.W: *The Electric Lines of the Lancashire & Yorkshire Railway*, Electric Railway Society, Sutton Coldfield, 1976

Gahan, J.W.: *Seaport to Seaside – lines to Southport and Ormskirk*, Countyvise, Birkenhead, 1985

Goslin, G.W.: *London's Elevated Electric Railway – the LB&SCR suburban overhead electrification, 1909-1929*, Connor & Butler, Colchester, 2002

Hall, S.: *Railway Detectives - 150 years of the Railway Inspectorate*, Ian Allan, London 1990

Henniker, Col. A.M. CBE, RE: *The Official History of the Great War — Transportation on the Western Front 1914-18*, HMSO, 1937

Heritage, T.R.: *The light track from Arras*, Plateway Press, Brighton, 1999

Hughes, G.: *Sir Nigel Gresley – The Engineer and His Family*, Oakwood Press, Usk, 2001

Humphry Ward, M.: *England's Effort*, Charles Scribner's Sons, New York, 1916

Hunt, D.: *The Tunnel — The Story of the Channel Tunnel 1802-1994*, Images Publishing 1994

Jackson, D.: *J.G. Robinson – A Lifetime's Work*, Oakwood Press, Headington, 1996

Jenkin, C.F.: *Single Phase Electric Traction*, Paper No.3647, Proc. I.C.E., Vol.CLXVII, 1907

Kennedy, Sir Alexander B.W.: *Report of the Electrification of Railways Advisory Committee*, HMSO, London, 1921

Kitson-Clark, Col. E.: *Kitsons of Leeds*, Locomotive Publishing Co., London, 1937

Lewis, B.: *The Cabry Family – Railway Engineers*, Railway & Canal Historical Society, 1994

Lydall, F.: *Motor and Control Equipments for Electric Locomotives*, I.E.E. Journal, Vol.52, No.230, 1914

Lyell, Col. D.: *The work done by Railway Troops in France during 1914-19*, Paper No.4335, Proc. I.C.E., Vol.CCX, 1920

MacNeill, M.: *Máire Rua – Lady of Leamaneh*, Ballinakella Press, Whitegate, Co Clare, 1990

McGee, P.A.: *The possibilities of employing Electric Motive Power on Irish Railways*, Trans. I.C.E.I., Vol. XLVI, 1921

Marshall, J.: *The Lancashire and Yorkshire Railway - Vol. 2*, David & Charles, Newton Abbot, 1970

Marshall, J.: *The Lancashire and Yorkshire Railway - Vol. 3*, David & Charles, Newton Abbot, 1972

Marshall, J.: *George Hughes of Horwich*, British Railway Journal, Vol.7, Nos.59, 60 & 61

Mason, E.: *The Lancashire and Yorkshire Railway in the Twentieth Century*, (2nd edition.) Ian Allan, London, 1961

Mason, E. (as 'Rivington'): *My Life with Locomotives*, Ian Allan, London, 1962

Massey, W.T.: *How Jerusalem Was Won — The record of Allenby's campaign in Palestine*, London, 1919

Neele, G.P.: *Railway Reminiscences*, McCorquodale & Co., London, 1904

Newham, A.T.: *The Bessbrook & Newry Tramway*, Oakwood Press, Blandford, 1979

227

Nock, O.S.: *The Lancashire and Yorkshire Railway — A Concise History*, Ian Allan, London, 1969

O'Brien, Hon. G: *These my Friends & Forebears – The O'Briens of Dromoland*, Ballinakella Press, Whitegate, Co. Clare, 1991

O'Brien, H.E.: *Report on Electric Traction in the U.S.A. with special reference to railroads operating with the system known as the third rail system*, Lancashire & Yorkshire Railway, Manchester, 1903

O'Brien, H.E.: *Electric Traction on Urban and Inter-urban Steam Railways*, Proc. Liverpool Engineering Society, April 1908

O'Brien, H.E.: *Improvements in Buffering Devices for use on Railways*, British Patent Specification No.19,362 of 1909

O'Brien, H.E.: *Continental Engineering*, Presidential Address to Engineering & Scientific Club of the Horwich Railway Mechanics Institute, November 1910

O'Brien, H.E.: *More Continental Engineering*, Presidential Address to Engineering & Scientific Club of the Horwich Railway Mechanics Institute, November 1912

O'Brien, H.E.: *The design of Rolling Stock for Electric Railways*, I.E.E. Journal, Vol.52, No.231, 1914 (Also published in Proceedings of the Engineering & Scientific Club of the Horwich Railway Mechanics Institute, Vol. IX, 1914)

O'Brien, H.E.: *The Federal Solution for Ireland*, Overseas Magazine, Vol. IV, No.41, June 1919

O'Brien, H.E.: *The Application of Electric Locomotives to Main Line Traction on Railways*, I.E.E. Journal, Vol.58, No.295, 1920

O'Brien, H.E.: *The Management of a Locomotive Repair Shop*, Paper No.86, I.Loco.E. Journal No.45, Vol. X, 1920

O'Brien, H.E.: *The Future of Main Line Electrification on British Railways*, I.E.E. Journal, Vol.62, No.333, 1924; and I.E.E. Journal, Vol.62, No.335, 1924

O'Brien, H.E.: *Main Line Electrification*, Paper No.175, I.Loco.E. Journal No.68, Vol. XV, 1924

O'Brien, H.E.: *The economic aspects of Railway Electrification*, Paper read at the 11th session of the International Railway Congress Association, Madrid, 1930

O'Brien, H.E.: *Irish Centre: Chairman's Address*, I.E.E. Journal, Vol.70, No.421, 1932

O'Brien, H.E.: *The growth of Electricity Supply and its relations to civilisation*, Paper read before the Statistical & Social Inquiry Society of Ireland, May 1933

O'Brien, H.E. & Gatwood W.: *Improvements in Self-contained Spring Buffers for Railway Vehicles and the like*, British Patent Specification No.6136 of 1911

O'Donoghue, J.: *Historical Memoir of the O'Briens*, Hodges Smith, Dublin, 1860, reprinted by Martin Breen, Ruan, Co.Clare, 2002

Öfverholm, I: *The Economic Problems of the Swedish State Railway Electrification*, I.E.E. Journal, Vol.72, No.437, 1933

Parris, H.: *Government and the Railways in Nineteenth-Century Britain*, Routledge & Keegan Paul, London, 1965

Pearson, P.: *Between the Mountains and the Sea*, O'Brien Press, Dublin, 1998

Pringle, Col. Sir John W.: *Report of the Railway Electrification Committee*, HMSO, London, 1928

Pritchard, Maj-Gen. H.L. (Ed.): *The History of the Corps of Royal Engineers*, Vols. V & VI, The Institution of Royal Engineers, Chatham, 1952

Richards, H.W.H: *Twelve Years' Operations of Electric Traction on the London Brighton & South Coast Railway*, Paper No.4441, Proc. I.C.E., Vol.CCXV, 1923

Rider, J.H.: *The Electrical System of the London County Council Tramways*, I.E.E. Journal, Vol.43, No.197, 1909

Rowledge, J.W.P.: *L&YR Express Engines*, Locomotives Illustrated No.88, March-April 1993

Rowledge, J.W.P.: *LMS Pre-Grouping Eight-Coupled Locomotives*, Locomotives Illustrated, No.107, May-June 1996

Rowledge, J.W.P.: *L&YR Passenger Tank Locomotives*, Locomotives Illustrated, No113, May-June 1997

Rowledge, J.W.P.: *L&YR Express 0-6-0 and 0-6-0ST Locomotives*, Locomotives Illustrated, No131, May-June 2000

Rush, R.W.: *Lancashire & Yorkshire Passenger Stock*, Oakwood Press, Lingfield, 1984

Rutherford, M: *Comparison and Revision: The Grouping and Early LMS Locomotive Policy*, Backtrack, Vol.19, No.3, March 2005

Schoepf, T.H.: *Single-phase Railway Motors and methods of controlling them*, I.E.E. Journal, Vol.36, No.178, 1906

Shepherd, E. & Beesley, G.: *The Dublin & South Eastern Railway*, Midland Publishing, Leicester, 1998

Simmons, J.: *The Express Train and other Railway Studies*, Thomas & Lochar, Nairn, 1994

Smith, M.D.: *Horwich Locomotive Works*, Wyre Publishing, St. Michael's on Wyre, 1996

Smith, R.T.: *Some Railway conditions governing Electrification*, I.E.E. Journal, Vol.52, No.228, 1914

Taylorson, K.: *The Narrow Gauge at War*, Plateway Press, Croydon, 1987

Taylorson, K.: *The Narrow Gauge at War 2*, Plateway Press, Brighton, 1996

Tomlinson, W.W.: *The North Eastern Railway — its rise and development*, Andrew Reid & Co., Newcastle and Longmans Green & Co., London, 1915

Topping, B.J.: *The Horwich 'Crabs'*, Steam Days, No.172, December 2003

Tufnell, R.M.: *The Unfortunate Case of Henry O'Brien*, Backtrack, Vol.13, No.2, February 1999

Warder, S.B.: *Electric Traction prospects for British Railways*, Paper No.498, I.Loco.E. Journal No.219, Vol. XLIV, 1951

Weir, Lord of Eastwood; Wedgwood, Sir Ralph; & McLintock, Sir William: *Report of the Committee on Main Line Railway Electrification*, HMSO, London, 1931

Wells, J.: *Horwich and the Labour Dispute of 1911*, Backtrack, Vol.19, No.1, January 2005

Whitaker, F.P.: *Rotary Converters, with special reference to Railway Electrification*, I.E.E. Journal, Vol.60, No.309, 1922

Newspapers and Periodicals

The Belfast Newsletter, 20th November 1835

The Economist, 15th February 1845

The Engineer: *A National Projectile Factory*, Vol.CXXII, 21st & 28th July 1916

Herepath, J.: Railway Magazine and Steam Navigation Journal, Vol.VI, No. XLI, July 1839

Horwich & Westhoughton Journal, No.2074, 7th October 1911

Horwich & Westhoughton Journal, No.4995, 29th September 1967

Horwich & Westhoughton Journal, No.5000, 3rd November 1967

Journal of the Irish Railway Record Society, *The arrest of William Smith O'Brien*, Vol.4, No.16, Spring 1955

Railway Gazette, *Report on I.Mech.E. Summer Meeting 1909*, Vol.47, No.6, 6th August 1909

Railway Gazette, *Special War Transportation Number*, 21st September 1920

Railway Magazine, *Electric Traction on the Lancashire & Yorkshire Railway*, Vol.XIV, No.80, February 1904

Railway Magazine, *Lancashire & Yorkshire Railway Special Is*sue, Vol.XXVIII, No.164, February 1911

Railway Magazine, *The Horwich Works of the Lancashire & Yorkshire Railway*, Vol. XXXIII. No.194, August 1913

Index

Personalities

Family

Bassett, Martha 29
Bellingham, Sir Alan Edward, 3rd Bart. Bellingham 39
Bellingham, Augusta Mary Monica (Marchioness of Bute) 83
Bellingham, Frances Anne Jane (H.E. O'Brien's mother-in-law) 39, **81**, 204
Bellingham, Sir Henry 81
Brian Boru (High King of Ireland) **9**, 11, 32, 47, 205
Burgoyne, Margaretta (Capt. William O'Brien's second wife) 20, 24
Byrne deSatur, Stella Alice Pauline (C.H. Merz's wife) 112, 204
Callander, Mary 16
Callander, Fanny 16
Callander, Col. James 16
Cooke, Frances Barbara 39, 204
Cooke, Revd Richard 39
Daly, Katherine 28
Falkiner, Irene 80, **81**
Finucane, Louisa 24
Fitzgerald, Anne 29
Fitzgerald, Mary 29
Fitzgerald, William 29
French, Anne 24
French, Lady Anne 81
French, Robert 24
Gabbett, Joseph 25
Gabbett, Lucy Caroline (wife of William Smith O'Brien) 25
Godley, Dennis 34
Godley, Harriet (H.E. O'Brien's grandmother) 28, 29, 31
Godley, Revd James 30
Godley, John 28
Godley, Sir John Arthur, 1st Baron Kilbracken 28
Godley, John Robert 28
Graham, Sir James Robert George 16, 17, 18
Gray, Elizabeth (née Deane) 15
Hamilton, Lucia 15
Henn, Elizabeth 16
Henn, Mary 16
Henn, William 16
Hickman, Mary 15
Hyde, Lady Anne 15
Hyde, Lady Frances 15
Keightley, Catherine 15
Keightley, Thomas 15
Lamb, Mary Oclanis 29
Lloyd, Rev. Richard 25
Massy-Dawson, Louisa 29
Maunsell, Lucy 25
Maunsell, Revd Richard 16
Maunsell, Ven. William Thomas 25
McCleverty, Elizabeth (wife of Capt. Donatus O'Brien) 17, 19
McKerrell, Maj. Reginald L'Estrange 29
Merz, Charles Hesterman 112, 169, 188, 199, **203**, 204
Merz, Pauline Barbara (daughter of C.H. Merz) 204
Merz, Robert deSatur (son of C.H. Merz) 204
O'Brien, Angelina Rose Geraldine (H.E. O'Brien's aunt) 29, 30, 41, 43, 109
O'Brien, Annabella Charlotte (H.E. O'Brien's aunt) 29
O'Brien, Augustus Stafford 15, 16
O'Brien, Brian Eoghan (H.E. O'Brien's son) 89, **90**, 96, 98, 100, 106, **113**, 119, **130**, 201, 203, 204
O'Brien, Charlotte Anne (daughter of Sir Lucius, 13th Baron) 24, 29
O'Brien, Conor (1582-1603) 12, 13
O'Brien, Conor (1617-1651) 12, 13
O'Brien, Dermod, 2nd Baron Inchiquin 10, 13
O'Brien, Sir Donat, 1st Bart. Dromoland 10, 13, 14, 15
O'Brien, Capt. Donatus, (General Secretary of the Railway Board) 7, 16, 17, 18, 19
O'Brien, Donough (son of Murrough, 1st Baron Inchiquin) 12, 13
O'Brien, Donough (second son of Sir Edward, 2nd Bart.) 15, 16
O'Brien, Sir Donough (grandfather of Sir Donat, 1st Bart.) 12, 13
O'Brien, Donough Ramhar 8, 13
O'Brien, Donough Cairbreach, King of Thomond (1210-42) 10, 11
O'Brien, Sir Edward, 2nd Bart. Dromoland 13, 14, 15, 16
O'Brien, Sir Edward, 4th Bart. Dromoland 13, 14, 15, 16, 24, 25, 28

O'Brien, Edward (brother of Rev. Henry) 29
O'Brien, Edward (H.E. O'Brien's uncle) 29
O'Brien, Sir Edward Donough, 14th Baron Inchiquin 27
O'Brien, Ellen (née Waller, H.E. O'Brien's mother) 32, 33, **34**, 38, 40, 41, **45**, 61, 68, 78, 80, 85, 88, 89, 90, 96, **97**, 101, 105, 108,109, **113**, 114, 119, 142
O'Brien, Grace Amy Frances (H.E. O'Brien's aunt) 29
O'Brien, Henry, 8th Earl of Thomond 13, 14
O'Brien, Henry (of Stonehall) 15
O'Brien, Hon. & Revd Henry (H.E. O'Brien's grandfather) 7, 24, 26, 27, 28, 29,30, 31
O'Brien, James, 3rd Marquis of Thomond 13, 14
O'Brien, James, MP (Capt. William's father-in-law) 13, 19
O'Brien, James, 7th Earl of Inchiquin 13, 24
O'Brien, Juliana (daughter of Sir Lucius, 13th Baron) 24
O'Brien, Katherine (H.E. O'Brien's aunt) 29, 41, 43, 79, 111
O'Brien, Katherine Lucy (wife of Capt. William) 19, 20
O'Brien, Lucius (1675-1717) (son of Sir Donat, 1st Bart.) 13, 14
O'Brien, Lucius (father of Capt. Donatus & Capt. William) 16, 19,
O'Brien, Sir Lucius, 13th Baron Inchiquin 13, 14, 24, 26, 27, 29,
O'Brien, Sir Lucius Henry, 3rd Bart. Dromoland 15, 16, 24
O'Brien, Mary Grace (daughter of Sir Lucius, 13th Baron) 24
O'Brien, Murrough, 1st Baron Inchiquin 10, 13
O'Brien, Murrough, 1st Earl of Inchiquin 13, 14
O'Brien, Murrough, 1st Marquis of Thomond 13, 14
O'Brien, Murrough John (H.E. O'Brien's father) 29, 31, 32, 33, **34**, 36, 38, 41, 42, 44, 48, 55, 61, 68, 72, 79, 80, **81**, 85, 88, 89, 92, 97, 101, 105, 108,112, 113, 114

O'Brien, Olivia Fiona (H.E. O'Brien's grand-daughter) 3, 7, 201
O'Brien, Olivia Henrietta (daughter of Rev. Henry) 29,30
O'Brien, Robert (brother of Hon. & Revd Henry) 26
O'Brien, Stafford 15
O'Brien, Capt. William (General Manager NER) 7, 16, 18, 19, **20**, 21, 22, 23
O'Brien, William Smith (brother of Hon. & Revd Henry) 24, 25, **26**, 27, 31
Smith, Charlotte (wife of Sir Edward O'Brien, 4th Bart) 16, 24, 25, 28
Smith, William (of Cahermoyle) 24
Smyth, John Watt 29
Smythe, Eileen Barbara (H.E. O'Brien's sister-in-law) 79, 80, **81**, **83**, 112
Smythe, Erica (niece of Frances Victoria Lucy) 80, **81**
Smythe Frances Isabella Anne (H.E. O'Brien's wife's aunt) 204
Smythe, Frances Victoria Lucy (H.E. O'Brien's wife) 3, 39, 68, 79, 80, **81**, 82, **83**, 85, 88, 89, **90**, 91, 96, 98, 100, 104, 105, 109, 112, **113**, 119, 187, 201, 204, 206
Smythe, Henry Meade 39, 204
Smythe, Olive Mary (H.E. O'Brien's sister-in-law) 68, 80, **81**,
Smythe, Capt. Richard Altamont (H.E. O'Brien's father-in-law) 39, 80, **81**, 204
Smythe, Rupert Caesar (H.E. O'Brien's brother-in law) 96, 97
Stafford, Susannah 15
Strahan, Elizabeth (H.E. O'Brien's daughter-in-law) 201
Toulmin, Revd Frederick Bransbury 29
Waller, Bolton 16, 32
Waller, Ellen (see O'Brien, Ellen)
Waller, John 32
Waller, John Thomas 16
Waller, Revd William 32
White, Hon. Ellen Harriet 27
White, Sir Luke, 2nd Baron Annaly 27
Wilson, John 29
Wilson, Col. John Gerald 29, **30**
Wilson, Sir Murrough John 30, 32, 44, 128, 187, **196**
Wilson, Richard Bassett (1806-67) 29
Wilson, Lieut-Col. Richard Bassett 30
Wilson, Lieut. Richard Bassett 30, 41

Military Officers

Allenby, Gen. Sir Edmund 146, 149, 150, 151, 152
Bodwell, Lieut-Col. Howard Lionel 146
Campion, Lieut. Arthur Havard Montriou **120**, **121**, 122
Chetwode, Lieut-Gen. Sir Philip 149
Cox, Brig-Gen. Charles Frederick 154
Fay, Lieut. Samuel Ernest **120**, **121**, 122
Haig, Field Marshal Sir Douglas 142, 144, 145
Harrisson, Brig-Gen. Geoffrey Harnett 151
Hodgins, Lieut-Col. Arthur Edward 146
Jordan-Bell, Maj. Charles Edward **121**, 123, 152, 215
Lefevre, Lieut-Col. Alfred George Tully 146
Logan, Lieut-Col. Malcolm Hunter 146
Mance, Brig-Gen. Sir Henry Osborne 144, 145
Myddleton, Lieut-Col. Cornelius William 146
Murray, Gen. Sir Archibald 149
Nash, Brig-Gen. Sir Philip Arthur Manley 144, 145
Nimmo, Lieut. James Valence **120**, **121**, 122, 125
Phillips, Lieut. George William **120**, **121**, 122, 123, 124
Stewart, Brig-Gen. John William 150
Stobart, Lieut-Col. Hugh Morton 151
Twining, Maj-Gen. Sir Philip Geoffrey 145, 146

Naval Officers

Dunkerley, Lt-Cdr. William Donald 203
Talbot, Lt. Francis Robert Cecil 203
Morris, Lt. Fenton Harry 203
Newell, Lt.(RNR) Daniel Edward Treymain 203

Parliamentarians & Diplomats

Addison, Christopher 127, 128, 129, 130, 131, 158
Aiken, Francis (Frank) Thomas (Irish Minister for Co-ordination of Defensive Measures, 1939-45) 204
Asquith, Rt.Hon. Herbert 42, 127
Baldwin, Rt.Hon. Stanley 194, 196
Blair, Rt.Hon. Tony 208
Bull, Sir William 194
Casey, Richard Gardiner 194
Cosgrave, William T. (President of the IFS Executive Council) 193
de Broqueville, Baron Charles (Belgian Prime Minister) 137
Geddes, Sir Eric Campbell 128, 142, 143, 145, 163, 175
George, Rt.Hon. Lloyd 127, 128, 129, 135, 137, 141, 142, 143, 149, 158
Gladstone, Rt.Hon. William 17, 29, 31, 33, 38
Hempel, Dr Eduard (Nazi German Minister to Ireland) 206
Kitchener, Lord Herbert 141, 143
Maffey, Sir John (British Representative to Ireland) 205
McDonald, Rt.Hon. Ramsey 194, 196
McGilligan, Patrick (Irish Minister for Industry & Commerce) 190, 191, 192, 198
Morley, Viscount John (Chief Secretary for Ireland, 1892-95) 37, 38, 55
Peacock, Sir Edward 194
Peel, Sir Robert 17, 19
Thomas, Sir Robert 194
Thurtle, Ernest
Wood, Rt.Hon. Thomas McKinnon 160

Public Servants

Coyne, Thomas Joseph 204, 205, 206
Cummins, G.C. (Board of Trade) 107
Ramsay, James Andrew Broun, 10th Earl of Dalhousie 17, 18, 19
Evans, Sir Worthington 159
Flynn, Thomas Joseph 164
Ingram, Joseph (formerly with Irish Railway Clearing House) 164, 167
Jebb, Lieut-Col. Sir Joshua 19
Laing, Samuel 17, 19
McDunphy, Michael 205, 206
Porter, George Richardson 17, 19
Wharton, Percy (formerly with GS&WR) 164

Railway Engineers

Anderson, James Edward (MR / LMS) 182, 183
Arter, William (L&YR) 57, 89, 92, 199
Ash, William James (MGWR / GSR) 191

Aspinall, Sir John Audley Frederick (GS&WR, L&YR) 4, 5, 55, 56, 60, 61, 63, 68, 71, 82, 95, 96, 100, 101, 103, 108, 110, **111**, 112, 113, 161, 163, 165, 172, 188, 201, 209

Attock, Frederick (L&YR) 60, 96

Attock, Frederick William (L&YR / LMS) 71, 83, 109, 111, 174, 182, 209

Bachellery, André (Chief Engineer, Chemins de fer du Midi) 180

Banks, George (L&YR) 60, 63, 209

Barnes, William Alfred (L&YR) 57, 58

Barnes, Frederick Stanton (L&YR) 83, **111**, 209

Bazin, John Ralph (GNR, GS&WR) 46,189, 190

Beames, Hewitt Pearson Montague (L&NWR / LMS) **121**, 123, 124

Beesley, Frederick Walter (CR / LMS) 6

Billington, John Robert (L&YR / L&NWR / LMS) 57, 58, 71, 109, **111**, 172, 174, 177, 184, 186,

Bond, Roland Curling (LMS / BR) 174

Bredin, Edgar Craven (GS&WR / GSR / CIE) 206

Cabry, Thomas (NER) 23

Chalmers, Walter (NBR) 6,174

Churchward, George Jackson (GWR) 170

Coey, Robert (GS&WR) 37, 50, 51, 55, 100

Cortez-Leigh, Frederick Augustus (L&NWR / LMS) 184

Cox, Ernest Stewart (L&YR / L&NWR/ LMS / BR) 176

Creagh, Harold (L&YR) 109, 111, 209

Cronin, Richard (DW&WR / D&SER) 50, 51, 52, 55, 71, 88, 100, 169

Crouch, John Peachy (L&YR) 57, 62, 82, 95, 98, 109, **111**, 209

Dalziel, James (MR / LMS) 92,178, 184

Davies, James (L&YR) 60

Dawson Sir Philip (LB&SCR) 104, 169,

Deeley, Richard Mountford (MR) 182, 183

Field, Basil Kingsford (LB&SCR) 174

Firth, Harold William (GER) 115, 116

Flamme, Jean Baptiste 112,113

Fowler, Sir Henry (MR / LMS) 4, 37, 57, 58, **111**, 128, 129, 141, 183, 186, 188, 195

Gobey, Francis Edward (L&YR) 63, 96, 98,**111**, 209

Gresley, Sir Herbert Nigel (L&YR, GNR / L&NER) 4, 55, 58, 60, 62, 63,71, 82, **111**, 184, 186,187, 188,196, 209

Grierson, Thomas Benjamin (DW&WR, LD&ECR) 48, 49, 50, 51

Haigh, James Harold (L&YR) 57, **111**, 174,

Harrison, Thomas Elliot (L&YR) 20

Housley, Harold (L&YR) 109, 111, 209

Howarth, James (L&YR) 60, 63, 209

Hoy, Henry Albert (L&YR) 56, 60, 62, 63, 70, 71, 77, 79, 82, 209

Hughes, George (L&YR / L&NWR / LMS) 7, 60, 62, 71, 73, 86, 89, 92, 96, 100, 101, 102, 103, 106, 107, 108 110, **111**, 113, 133, 170, 172, 175, 176,177, 179, 180, 181, 182, 183,184,185, 186, 209

Humphrey, Barnard (GWR) 58

Hutchinson, Thomas Cross (L&YR) 57

Ivatt, Henry Alfred (GS&WR, GNR) 25

Jones, Arthur Dansey (L&YR, SE&CR) 60, 62, 91, 109, 111, 209

Jones, Herbert (L&SWR / SR) 164

Kempt, Irvine (CR) 6,174

Lund, Arthur (L&YR) 111, 117, 209

Mackay, Charles O'Keeffe (L&YR) 60, 62, 209

Maunsell, Richard Edward Lloyd (GS&WR, SE&CR / SR) 4, 16, 24, 25, 33, 50, **51**, 55, 111, 161, 197

McLaughlin, Reginald George (L&YR / LMS) 57, 98, **111**, 209

Montgomery, Charles Hubert (L&YR) 62, 71, 95, 96, 109, 111, 209

Morgan, Lieut-Col. Sir Charles (Chief Engineer LB&SCR) 148

Morton, William Herbert (MGWR / GSR) 4, 7, 46, **47**, 161, 190, 191, 199, 200, 202, 203

Parr, William Henry Marsh (L&YR) 57, 58

Parry, Joseph (GCR) 174

Pickersgill, William (CR) 6,

Ramsbottom, John (L&NWR, L&YR) 55, 56

Raven, Sir Vincent Litchfield (NER) 128, 160, 187, 220

Richards, Henry Walter Huntingford (LB&SCR, L&NER) 164, 180, 187

Riches, Tom Hurry (TVR) 166

Roberts, Gervase Henry (L&YR) 57, 60, 62, 109, 111, 160, 161, 209

234

Robinson, John George (WL&WR, GCR) 47, 48, 167
Sayers, Josiah (MR / LMS) 92,
Shawcross, George Nuttal (L&YR / L&NWR / LMS) 57, 107, **111**, 119, 165, 186
Simpson, Col. Lightly Stapleton, CBE, DSO 150, 163
Smith, David (G&SWR) 6
Smith, Roger Thomas (GWR), I.E.E. President 114, 177, 178
Stanier, Sir William Arthur (GWR, LMS) 200
Storey, Warren (GS&WR / GSR) 190, 191, 202
Symes, Sandham John (GS&WR, MR / LMS) 37
Tetlow, Zechariah (L&YR) 109, 172
Wakefield, William (DW&WR) 49
Watson, Edward Abraham Augustus (GS&WR) 167
Whitelegg, Robert Harben (G&SWR, LT&SR) 6
Wild, George Henry (D&SER) 4, 169
Winder, Oliver (L&YR) 57, 60, 62, 63, 71, 82, 85, 86, 91, 95, **111**, 129, 131, 209
Worthington, William Barton (L&YR) 62
Wright, William Barton (L&YR) 55, 56

Railway Inspecting Officers

Coddington ,Capt. Joshua 17
Druitt, Maj. Edward 70, 85, 93, 94
Ford, Alfred Montague 54
Hutchinson, Maj-Gen. Charles Scrope, RE, CB 22
Pasley, Gen. Charles William 17, 18, 19
Pringle, Col. Sir John Wallace 187
Tyler, Capt. Sir Henry Whatley 22,23
Yolland, Col. William 22, 23

Railway Officials

Armytage, Sir George John (L&YR) 56, 95
Bayley, Charles William (L&YR) 62, 98
Burgess, Henry Givens (DW&WR / D&SER, L&NWR / LMS) 164, **165**, 181
Calthrop, Sir Guy (L&NWR) 148
Cawkwell, William (L&NWR) 21
Crompton, James Hamer (L&YR) 159

Cusack, Sir Ralph Smith (MGWR) 37
Dale, Frederick Thorpe (L&YR) 63
Fielden, Samuel (L&YR) 56
Follows, John Henry (MR / LMS) 180
Galloway, Col. William Johnson (L&NER) 197
Granet, Sir William Guy (MR / LMS) 145, 181, **182**, 183
Grey, Edward, Viscount Grey of Fallodon (L&NER) 197
Hauxwell, Samuel (L&YR) 62
Henderson, Sir Alexander (Lord Farringdon, L&NER) 197
Leeman, George (NER) 21
Maunsell, Edward William (L&ER, DW&WR) 24, 25
Neele, George Potter (L&NWR) 21, 22
Nugent, Sir Walter (MGWR / GSR) 190
Ormsby, Francis Balfour (GS&WR) 39
Ponsonby, John William, 4th Earl of Bessborough (GWR) 17
Roche, Sir David (L&ER) 24
Saunders, Charles (GWR) 20
Symes, Sandham (GS&WR) 37
Tatlow, James Teare (L&YR) 56, 62, 63, **111**
Thompson, Henry Stephen (NER) 23
Thompson, Yates (L&YR) 55
Watson, Sir Arthur (L&YR / L&NWR / LMS) 175, 181
Wedgwood, Sir Ralph Lewis (NER / L&NER) 128, 145, 187
Wharton, Josiah (L&YR) 63

Railway Staff

Boote, William (signalman L&YR) 84, 85
Clayton, Thomas (shop foreman L&YR) 60, **102**
Grime, John Townend (accountant L&YR) 99
Hulme, Billy (guard GS&WR) 26
Lloyd, William (engine driver L&YR) 70
Mellin, Ventry Guiscard 80, **81**
O'Leary, Arthur (clerk GS&WR) 26
Pilling, Philip (assistant accountant L&YR) 99
Rigby, Edward (fireman L&YR) 69
Rimmer, William (motorman L&YR) 84, 85
Smith, Stuart Mansel (accountant L&YR) 99

Wadsworth, George Wilfred (draughtsman L&YR) 102

Royalty & Heads of State

Duke of Edinburgh 208
Hyde, Douglas (President of Ireland) 204
King George V 156
King Henry VIII 10
King James II 15
Mitterrand, François (President of France) 208
Queen Anne 15
Queen Elizabeth II 208
Queen Mary II 15

Others

Aspinall, Edith (daughter of J.A.F.Aspinall) 60, 63
Aspinall, Isabel (daughter of J.A.F.Aspinall) 60
Austin, James Valentine (Judge) 107
Beharrell, Sir (John) George 144, 145
Bell, Revd Joseph Samuel 37
Brighouse, Samuel (Coroner) 70
Brocklebank, Thomas Haslehurst 46
Brotherhood, Alfred (Electrical Engineers Institute, New York) 64, 66
Carter, Frederick William (British Thomson-Houston Co.) 110, 178
Clements, Robert ('Bob') Nathaniel 25, 197
Collard, William 195, 196
Colles, Graves Chamney 36, 37
Colles, William 36
Deane, Thomas Newenham (architect) 39
Debauche, Hubert 137
Décauville, Paul 146
Dendy Marshall, Chapman Frederick 195
d'Erlanger, Baron Emile 194, 195
Donaldson, John Muir (President of I.E.E.) 199
Drumm, Dr James Joseph 190
Ellsworth, Lincoln 193
Exham, William 39
Fay, James Matthew 190
Fielden, Sarah Jane (wife of Samuel) 56
Fletcher, Revd Dudley 80
Furnivall, Willoughby Charles 49
Gatwood ,Walter 99
Girouard, Sir Percy 143
Gray, Robert Kaye 67

Harriss, George Marshall (DUTC) 199
Haslam, Sidney Bertram 166
Humphry-Ward, Mary (née Arnold) 135
Jenkin, Charles Frewin 87, 88
Kennedy, Sir Alexander Blackie William 164
Kitson, James 44, 45,46
Kitson, John Hawthorn 46
Kitson, James Jnr. 46
Kitson-Clark, Col. Edwin 182
Lever, Sir Hardman 158
Lydall, Francis (Merz & McLellan) 114, 178, 195, 196
Mann, Sir John 158
Maunsell, Daniel 24
Maunsell, George Meares 24
Maunsell, Dr Henry 24
Maunsell, Henry, JP 24
Maunsell, John, JP (father of R.E.L. Maunsell) 33, 36
Maunsell, Robert (R.E.L. Maunsell's grandfather) 24
McGee, Patrick A. 168
Nussey, Obidiah (woollen merchant) 45
Pain, James & George Richard (architects) 14
Parsons, Hon. Charles Algernon 45
Parsons, Richard Clere 45
Reay, Thomas Purvis 46
Rider, John Hall 94
Rolls, Charles Stewart 60
Sanderson, Richard Philip Charles 182
Schoepf, Theodore Hausmann 87
Sheldon, Samuel (Brooklyn Polytechnic Institute) 64
Smith, Philip 56
Stoney, Canon Robert Baker 80
Symes, Sandham (architect) 37
Trotter, Alexander Pelham (Electrical Adviser to the BoT) 93, 94
Vaughan, Edward Littleton (Eton schoolmaster) 41, 42, 43, 44, 105, **120**
Warren, Robert 35, 37
Weir, William Douglas (Lord Weir of Eastwood) 187
Whitaker, Frank Percy 175
Williams, William Cyril 182

Railway Accidents

Brockley Whins (NER) 22, 23
Cargo Fleet (NER) 22

Hall Road (L&YR) 83, **84**, 85, 86
Marsh Lane (L&YR) 93, 117
Scotswood Bridge (NER) 22
St. Nicholas Crossing (NER) 21, 23
Waterloo (L&YR) 68, **69**, 70
Wicklow Junction (DW&WR) 52, 53, **54**

Railway Companies

Albany & Hudson Electric Railway 66
Albany & Schenectady Railroad 65
Aurora, Elgin & Chicago Railroad 65
Baltimore & Ohio Railroad 128
Belfast & County Down Railway (B&CDR) 7, 170, **171**
Bessbrook & Newry Tramway 167,168
Boston Elevated Railway 66
Brooklyn Heights Railway 66
Buffalo & Lockport Railway 65
Caledonian Railway (CR) 6, 174
Central Argentine Railway 109
Central London Railway 62, 71
Chicago, Milwaukee & St. Paul Railroad 220
Córas Iompair Éireann (CIE) 6, 7, 52, 193, 206, 208, 231
Dublin & South Eastern Railway (D&SER) 4, 6, 52, 100, 164, 165, 169, 189, 231
Dublin & Wicklow Railway (D&WR) 35, 36
Dublin Wicklow & Wexford Railway (DW&WR) 7, 25, 33, 36, 40, 46, 48, 49, 50, 51, 52, 53, 55, 71, 88, 168, 169
Giant's Causeway Tramway 44
Glasgow & South Western Railway (G&SWR) 6,
Great Eastern Railway (GER) 115
Great North of England Railway (GNoER) 17, 19, 20
Great Northern Railway (GNR) 82, 122, 183, 188
Great Northern Railway (Ireland) (GNR(I)) 7, 61, 170, **171**,
Great Southern Railways (GSR) 4, 7, 46, 47, 51, 52, 161, **162**, 189, 190, 191, 192, 199, 202, 203
Great Southern & Western Railway (GS&WR) 25, 26, 37, 39, 50, 51, 55, 61, 100, 164, 166, 167
Great Western Railway (GWR) 17, 18, 20, 32, 37, 46, 58, 114, 170, 175, 183

Interborough Rapid Transit Company (NY City Subway Operator) 66
International Railway Company (Buffalo, NY) 65
Lake Shore & Michigan Southern Railway 65
Lancashire & Yorkshire Railway (L&YR) 4, **5**, 6, 7, 55, **56**, 57, 58, 59, 60, 62, 63, 64, 66, 67, 68, **69**, 71, 72, 76, **77**, **78**, **79**, 80, 82, 83, 86, 89, 90, 92, 95, 96, 98, 100, **101**, **103**, 107, 108, **109**, **110**, **111**, 112, 114, 115, 117, 126, 161, 162, 164, 165, 166, 169, 170, 171,172, 174, 175, 176, 209, 210, 211, 212, 213
Lancashire Derbyshire & East Coast Railway (LD&ECR) 49,51
Leeds Northern Railway 20
Limerick & Ennis Railway (L&ER) 24, 25, 48
Liverpool Overhead Railway (LOR) 71, 80, 83,
London & Brighton Railway 18
London & Croydon Railway 18
London & North Eastern Railway (L&NER) 122, 175, 183, 187, 188, 189, 196, **197**, **198**, 220
London & North Western Railway (L&NWR) 2, 4, 21, 32, 124, 148, 164, 165, 175, 176, 183, 184, **185**
London & Paris Railway 195
London Brighton & South Coast Railway (LB&SCR) 90, 104, 105, 148, 164, 174, 180
London Midland & Scottish Railway (LMS) 4, 5, 7, 32, 58, 118, 165, 175, 176, 178, 180, 181, 182, 183, 184, **185**, 186, 187, 188, 189, 195, 200, **201**, 207
Long Island Railroad 67
Manhattan Elevated Railway 64, 66
Metropolitan Railway 71
Metropolitan District Railway 71
Metropolitan Elevated Railway (Chicago) 65
Midland Railway (MR) 7, 21, 37, 90, 92, 128, 145, 178, 181, 182, 183, 186
Midland Great Western Railway (MGWR) 37, 46, 47, 161, **162**,
Newcastle & Berwick Railway 20
Newcastle & Carlisle Railway 21

Newcastle & Darlington Junction Railway 17, 20
New York, New Haven & Hartford Railroad 66
North British Railway (NBR) 6, 174
North Eastern Railway (NER) 5, 20, 21, 22, 23, 30
Northwestern Elevated Railroad (Chicago) 65
Pennsylvania Railroad 65
Rhondda & Swansea Bay Railway (R&SBR) 49
South Eastern Railway (SER) 18, 20
South Eastern & Chatham Railway (SE&CR) 25, 51, 161
Southern Railway (SR) 25, 51, 105, 187, 195
Stockton & Darlington Railway 21
Toledo & Western Railway 65
Waterford & Limerick Railway (W&LR) 47, 48
Waterford Limerick & Western Railway (WL&WR) **48**
Waterloo & City Railway 71
West Clare Railway (WCR) **27**
Wilts Somerset & Weymouth Railway 20
York & Newcastle Railway 20
York & North Midland Railway 20
York Newcastle & Berwick Railway 20

Railway Locomotive and Rolling Stock Manufacturers

American Locomotive Company (Schenectady) 66
Baldwin Locomotive Works (Philadelphia / Eddystone) 65, 182
Andrew Barclay (Kilmarnock) 129
Beyer Peacock & Company (Gorton, Manchester) 49, 71, 77, 129, 170, **171**, 182
Birmingham Railway Carriage & Wagon Company 129
Brush Electrical Engineering (Loughborough) 129
Davey Paxman Ltd (Colchester) 129
Dick, Kerr & Company (Preston) 63, 67, 76, 77, 86, 109, 110, 117, 131, 141, 210
Dübs & Company (Queen's Park Works, Glasgow) **27**
Hudswell-Clarke (Leeds) 129
Hunslet Engine Co. (Leeds) 129

Hurst Nelson (Motherwell) 129
Kitson & Company (Airdale Foundry, Leeds) 45, 46, **47**, **48**, 49, **51**, 58, 182
Manning Wardle (Leeds) 130
Midland Railway Carriage & Wagon Company (Birmingham) 130
Metropolitan Vickers (Trafford Park, Manchester) 196
Neilson & Company (Hyde Park Works, Glasgow) 49,
North British Locomotive Company (Glasgow) 37, 130
Robert Stephenson (Newcastle-upon-Tyne) 130
Sharp Stewart & Company (Atlas Works, Glasgow) 49,
The Vulcan Foundry (Newton-le-Willows, Lancashire) 49, 51

Railway Vehicles

Steam Locomotive Classes

B&CDR 4-4-2T 170, **171**
DW&WR Vulcan Foundry 4-4-0 49
DW&WR Kitson 0-6-2T 49, 50, **51**
DW&WR 2-4-2T **52**
DW&WR 0-6-0 (Altered 0-6-2T - GSR J1 class) 51, 52
GNR(I) U class 4-4-0 170, **171**
GWR Saint class 4-6-0 170
L&NER A1 class 4-6-2 186
L&NER A4 class 4-6-2 196, **197**, **198**
L&YR K2 class 2-4-2T 106, 212, 213
L&YR B7 class 0-4-0ST 106, 212, 213
L&YR Q3 class 0-8-0 106, 212, 213
L&YR F22 class 0-6-0 108, 212, 213
L&YR Q4 class 0-8-0 108, 113, 212, 213
L&YR F19 class 0-6-0 212, 213
L&YR N1 class 4-6-0 172, 176, **177**, 184, **185**, 186, 212, 213
L&YR Railmotor 108, 212, 213
LMS 2-6-6-2 (Garratt) 195
LMS 4-6-2 (Turbomotive) 200, **201**
MGWR D1 class 2-6-0 (GSR K1 class) 161, **162**
WCR 0-6-2T **27**
WL&WR Kitson 0-4-4T 47, **48**
WL&WR Kitson 4-4-2T 47, 48
WL&WR Kitson 4-4-0 47, 48
WL&WR Kitson 0-6-0 48

Electric Rolling Stock

CIE DART electric multiple units 207
GSR Drumm Battery prototype railcar 191
GSR Drumm Battery 2-car articulated units 192, 193, 202
L&YR Experimental electric locomotive 102, 103, 104
L&YR Holcombe Brook experimental stock 110, 118
L&YR Liverpool – Southport direct control stock 5, 76, 77, 78, 79, 84, 210
L&YR Liverpool – Southport multiple unit stock 91, 210
L&YR Manchester – Bury stock 117, 118, 210
L&YR Lightweight stock for LOR through services 86, 210

Railway Workshops

Broadstone (MGWR) 46, 161, 162
Grand Canal Street (DW&WR / D&SER) 48, 50, 52, 53
Horwich (L&YR) 4, 55, 56, 57, 59, 60, 61, 62, 65, 68, 71, 95, 98, 100, 101, 102, 104, 106, 107, 108, 109, 111, 118, 119, 126, 129, 138, 161, 164, 165, 170, 171, 172, 174, 176, 183, 184, 186
Inchicore (GS&WR / GSR / CIE) 25, 37, 50, 55, 56, 100, 191, 192, 206
Miles Platting (L&YR) 56, 60, 96
Newton Heath (L&YR) 60, 62, 63, 71, 77, 85, 86, 94, 95, 96, 97, 98, 99, 101, 117

Specific Railway Routes

Baltimore & Ohio Belt Line 64
Channel Tunnel Rail Link (CTRL) 208
Dublin Area Rapid Transit (DART) 36, 207, 208
Dublin (Amiens Street) - Bray 4, 52, 168, 169, 190, 192, 202, 203
Harcourt Street Line 36, 169, 192, 202,
Holcombe Brook branch 109, 110, 117, 118
Lancaster - Morecambe - Heysham 90, 92, 182, 189
Manchester - Bury 117, 118, 176, 181

General Index

Airships (R100 and R101) 118, 193
Áras an Uachtaráin 205, 206
Babcock & Wilcox 6, 117, 131, 136, 141
Board of Trade 17, 19, 21, 22, 23, 54, 67, 70, 85, 93, 107, 116
Celia Limited 190
Dublin United Tramways Company (DUTC) 52, 199
Formby Power House 72, 73, 74, 75, 86, 88, 89, 101, 102
General Electric Company 64, 65, 66
HM Submarine *Thames* 203
Institution of Civil Engineers (I.C.E.) 8, 48, 55, 58, 63, 87, 92, 104, 162, 180
Institution of Civil Engineers of Ireland (I.C.E.I.) 34, 168, 199
Institution of Electrical Engineers (I.E.E.) 8, 67, 179, 216
Institution of Engineers of Ireland (I.E.I.) 8, 231
Institution of Locomotive Engineers (I.Loco.E.) 6, 58, 165, 172, 174, 182, 183, 189
Institution of Mechanical Engineers (I.Mech.E.) 8, 58, 67, 95, 110, 166, 231
International Railway Congress Association 100, 187
Irish Centre Party 162, 163
Irish Hospitals' Sweepstakes 204
Irish Railway Record Society (IRRS) 7, 8, 197,
Merz & McLellan 112, 150, 178, 188, 189, 190, 200,
Ministry of Munitions 4, 117, 126, 127, 129, 133, 137, 141, 142, 145, 158, 159, 162, 183
Ministry of Transport 4, 7, 163, 164, 165, 169, 171, 172, 175, 189,
RMS *Celtic* 64
RMS *Titanic* 58
SS *Armenian* 66
Thomson Houston Company 66
Trinity College Dublin 8, 28, 30, 31, 190, 199
Victoria University (Leeds) 46, 47, 55
Westinghouse Air Brake Works (Wilmerding) 65
Westinghouse Electric Manufacturing Co. (Pittsburgh) 64, 65

Note on the Author

Gerald Beesley, who is in the fourth generation of a family line of engineers, served as Commissioner for Railway Regulation in Ireland (2010-17). He was educated at St. Andrew's College, Dublin; Rye Grammar School, Sussex; and the Brighton Polytechnic where he gained an honours BSc degree in mechanical engineering. He undertook basic practical engineering training with Kearney & Trecker, Brighton, and obtained experience in the design of internal-combustion engines under the late Aubrey Woods at Weslake & Co., Rye, and subsequently with him at Anglo American Racers, Ashford, Kent.

From 1970 until 1989 he held various engineering positions in both the railway and the road transport divisions of Córas Iompair Éireann (CIÉ) — the Irish Transport Company. In January 1987 he was seconded for two years as Assistant Chief Mechanical Engineer and Workshops Manager to the newly formed Botswana Railways where he was instrumental in the establishment and initial development of their mechanical engineering department at Mahalapye. In 1990 he moved to the private sector in the UK, where he was responsible for setting up the private-owner freight locomotive operation for ARC Southern at Whatley, Somerset, and subsequently for establishing National Power's open-access freight train operations at Ferrybridge, West Yorkshire; the latter under the directorship of the late Keith McNair. From 1996 he worked on railway projects in Pakistan, Georgia, Lithuania, Zimbabwe, Tanzania, Zambia and Botswana prior to his appointment as Commissioner in August 2010.

Gerald is a Chartered Engineer and a Fellow of the Institution of Mechanical Engineers, the Institution of Engineers of Ireland, the Institution of Railway Signal Engineers, the Permanent Way Institution and the Irish Academy of Engineering. He has a long-standing interest in the development of railway technology and safety, especially as applied to the railways in Ireland, and he is co-author with Ernie Shepherd of an illustrated history of the Dublin & South Eastern Railway (Midland Publishing, 1998).

Approaching Social Theory

E. Ellis Cashmore and
Bob Mullan

Cartoons by Mick Davis

Heinemann Educational Books

Acknowledgement

We would like to thank our colleagues, Colin Bell, Martin Hollis and Dave Podmore, for commenting on an earlier draft of this book.

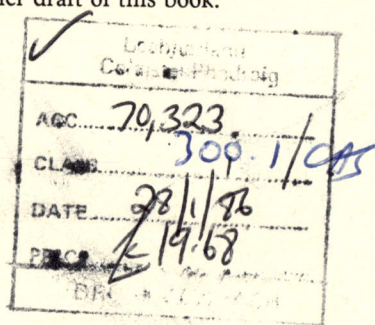

Heinemann Educational Books Ltd
22 Bedford Square, London WC1B 3HH
LONDON EDINBURGH MELBOURNE AUCKLAND
HONG KONG SINGAPORE KUALA LUMPUR NEW DELHI
IBADAN NAIROBI JOHANNESBURG
EXETER(NH) KINGSTON PORT OF SPAIN

© Ernest Cashmore and Bob Mullan 1983
First published 1983

British Library Cataloguing in Publication Data
Cashmore, E. Ellis
 Approaching social theory,
 1. Social sciences
 I. Title   II. Mullan, Bob
 300   HB85

 ISBN 0-435-82167-9
 ISBN 0-435-82168-7 Pbk

Phototypesetting by Georgia Origination, Liverpool
Printed in Great Britain by Biddles Ltd, Guildford, Surrey

Contents

Acknowledgement	*iv*
Introduction	*viii*

Part One: Behaviourism

1 **Man Machine** — 1
   Body and soul — 1
   Pavlov's dogs — 5
   Experiments with Albert — 9
   Changes from without — 13

2 **Plastic People** — 16
   The reinforcer — 16
   Inside the Skinner box — 22
   The technology of behaviour — 24
   Demons and goblins — 30

3 **Exchange and Reward** — 35
   In the market — 35
   Quid pro quo: Homans — 40
   Blau power — 42
   Individualism — 46

Theory into Action: Sex, Violence and the Media — *51*

Part Two: Interactionism

4 **Action Man** — 57
   The importance of meaning — 57

	A matter of interpretation: Weber	62
	Ironies	65
	The pragmatists	68
5	**The Selves**	76
	Explorations of consciousness	76
	All in the mind: Cooley	79
	The definition of the situation	84
	Mead's social theory of mind	87
6	**Society as Process**	97
	Disembodied society	97
	Roles and identities	103
	Back-stage with Goffman	108
	Society as a house of cards	113

Theory into Action: Sexual Deviance — 123

Part Three: Structuralism

7	**Social Wholes**	129
	Greater than the sum of the parts	129
	The organic model	133
	The anthropologists	140
	Building sight: Marx and Engels	144
8	**Psycho-structures**	154
	Of madness and majesty	154
	Schemes of thinking: Piaget	160
	The innate blueprint	165
	Universals	171
9	**Mistrust of Consciousness**	180
	Systems	180
	GST: the manic drive for unity	186
	The Two Marxisms	187
	Power-knowledge: Foucault	195

Theory into Action: The Capitalist State — 202
Epilogue — 205
Bibliography — 209
Index of Names — 225
Title Index — 230
Subject Index — 232